SpringerBriefs in Computer Science

SpringerBriefs present concise summaries of cutting-edge research and practical applications across a wide spectrum of fields. Featuring compact volumes of 50 to 125 pages, the series covers a range of content from professional to academic.

Typical topics might include:

- A timely report of state-of-the art analytical techniques
- A bridge between new research results, as published in journal articles, and a contextual literature review
- A snapshot of a hot or emerging topic
- An in-depth case study or clinical example
- A presentation of core concepts that students must understand in order to make independent contributions

Briefs allow authors to present their ideas and readers to absorb them with minimal time investment. Briefs will be published as part of Springer's eBook collection, with millions of users worldwide. In addition, Briefs will be available for individual print and electronic purchase. Briefs are characterized by fast, global electronic dissemination, standard publishing contracts, easy-to-use manuscript preparation and formatting guidelines, and expedited production schedules. We aim for publication 8–12 weeks after acceptance. Both solicited and unsolicited manuscripts are considered for publication in this series.

**Indexing: This series is indexed in Scopus, Ei-Compendex, and zbMATH **

Junmei Yao • Kaishun Wu

Cross-Technology Coexistence Design for Wireless Networks

 Springer

Junmei Yao ⓘ
College of Computer Science and Software
Engineering
Shenzhen University
Shenzhen, China

Kaishun Wu ⓘ
College of Computer Science and Software
Engineering
Shenzhen University
Shenzhen, China

ISSN 2191-5768   ISSN 2191-5776   (electronic)
SpringerBriefs in Computer Science
ISBN 978-981-99-1669-6   ISBN 978-981-99-1670-2   (eBook)
https://doi.org/10.1007/978-981-99-1670-2

This Springer imprint is published by the registered company Springer Nature Singapore Pte Ltd.
The registered company address is: 152 Beach Road, #21-01/04 Gateway East, Singapore 189721,
Singapore

Preface

With the rapid growth of Internet of Things (IoT), it is becoming a common phenomenon that numerous devices with different wireless technologies share the same band. WiFi and ZigBee are the two most important wireless technologies in IoT. WiFi is used for wireless local area networks (WLAN), while its market has stable increase now and in the future. ZigBee plays an important role in providing low cost, low data rate, and low energy consumption characteristics for wireless sensor networks. The ZigBee market also increases steadily these years. The rapid deployment of these heterogeneous devices leads to severe cross-technology coexistence issues, such as the cross-technology interference (CTI) management and cross-technology communication (CTC). The CTI issue arises from the fact that the IoT devices access the wireless channel independently in a random way, they have a high probability to interfere with each other since the heterogeneous IoT devices are not coordinated efficiently. The CTC issue exists as the heterogenous devices need to transmit information directly to benefit some applications, such as enabling the WiFi AP to directly control the ZigBee devices deployed for smart home.

In this book, we design a series of mechanisms to combat the cross-technology coexistence problem in heterogeneous wireless networks. In particular, we introduce background of heterogeneous wireless technologies as well as the cross-technology coexistence problem in details and further review the related literature. We further present a framework to describe the protocol design in this book. According to the framework, we first present a heterogeneous signal identification method based on both signal feature extraction and deep learning approach; this is the basis for the following protocol design. Then we present two cross-technology interference management mechanisms through the time-domain and frequency-domain system design, to improve the network performance. We also present a cross-technology communication mechanism through symbol-level energy modulation to boost new applications in IoT scenarios. Finally, we discuss some possible research directions, such as coordinated mechanism design for interference management, sensing-assisted cross-technology coexistence design, and waveform design through payload encoding, to further improve the efficiency of cross-technology coexistence.

The intended audience of this book shall be the readers, e.g., researchers, students, or even professionals, who are interested in the research areas of wireless networking, wireless communication, mobile computing, and IoT. In addition, this book can serve as a primer for beginners to gain a big picture of cross-technology coexistence in heterogeneous wireless networks; it could help them understand the key problems which affect the coexistence efficiency, and also the basic method of protocol design in wireless networks.

Shenzhen, China   Junmei Yao
February 2023   Kaishun Wu

Acknowledgments

This book was supported in part by the China NSFC Grants (62072317, 61872248, U2001207), Guangdong NSF 2017A030312008, Shenzhen Science and Technology Foundation (No. ZDSYS20190902092853047, R2020A045), the Project of DEGP (No.2019KCXTD005, 2021ZDZX1068), the Guangdong "Pearl River Talent Recruitment Program" under Grant 2019ZT08X603.

Contents

Acronyms

BPSK	Binary Phase Shift Keying
CCA	Clear Channel Assessment
CNN	Convolutional Neural Network
CSMA/CA	Carrier Sense Multiple Access/Collision Avoidance
CTC	Cross-Technology Communication
CTI	Cross-Technology Interference
FFT	Fast Fourier Transform
IoT	Internet of Things
ISM	Industrial, Scientific and Medical
MAC	Medium Access Control
OFDM	Orthogonal Frequency Division Multiplexing
OQPSK	Offset Quadrature Phase Shift Keying
PER	Packet Error Rate
QPSK	Quadrature Phase Shift Keying
RSSI	Received Signal Strength Indicator
SER	Symbol Error Rate
SINR	Signal to Interference plus Noise Ratio
SNR	Signal to Noise Ratio
USRP	Universal Software Radio Peripheral

Chapter 1
Introduction

Abstract In this chapter, we will introduce the background information of cross-technology coexistence, then review and discuss the related works on heterogeneous signal identification, cross-technology interference management and cross-technology communication. Finally, we will present the organization of this book.

Keywords WiFi · ZigBee · Cross-technology coexistence

1.1 Overview

With the widespread proliferation of the Internet of Things (IoT), it is becoming a common phenomenon that numerous devices with different wireless technologies (e.g., WiFi and ZigBee) share the unlicensed ISM spectrum. WiFi and ZigBee are the two most common wireless technologies in IoT. WiFi (IEEE 802.11) is used for wireless local area networks (WLAN), while its market has stable increase now and in the future. Cisco predicts that the number of WiFi hotspots will reach 628 Million by 2023 [9]. Meanwhile, ZigBee (IEEE 802.15.4) plays an important role in providing low cost, low data rate, and low energy consumption characteristics for wireless sensor networks. The ZigBee market also increases steadily these years. It was valued at USD 2.81 Billion in 2018 and is projected to reach USD 5.38 Billion by 2026 [34].

When both kinds of networks share the same unlicensed spectrum at the 2.4 GHz ISM (industrial, scientific and medical) band, the rapid deployment of heterogeneous devices leads to severe cross-technology coexistence problem, such as cross-technology interference management (CTI) and cross-technology communication (CTC). The CTI issue exists as the wireless devices access the wireless channel independently in a random way, they have a high probability to interfere with each other since the heterogeneous wireless devices are not coordinated efficiently. The CTC issue exists as the heterogeneous devices need to transmit information directly to benefit some applications, such as enabling the WiFi AP to directly control the ZigBee devices deployed for smart home. Meanwhile, signal identification is important for both CTI and CTC, as a device should obtain the

J. Yao, K. Wu, *Cross-Technology Coexistence Design for Wireless Networks*,
SpringerBriefs in Computer Science, https://doi.org/10.1007/978-981-99-1670-2_1

heterogeneous signal type and channel at first to better combat the two problems. In this book, we will provide the background information, review and summarize the recent advances, introduce several novel mechanisms to combat the problems, and finally present our outlooks on this research domain.

1.2 Background

In this part, we introduce the WiFi and ZigBee networks, especially their differences in both the physical and MAC (Medium Access Control) layers, and finally illustrate the cross-technology coexistence problem through the smart home scenario.

1.2.1 Differences of WiFi and ZigBee

1.2.1.1 The Channel Specifications

The WiFi and ZigBee devices can work on different frequency spectrums according to the standards. WiFi can work on both 2.4 and 5 GHz frequency bands. Although IEEE 802.11ac (called WiFi 5) [20] only supports the 5 GHz band, the newest standard IEEE 802.11ax (called WiFi 6) [22] still supports dual frequency bands due to the larger coverage of 2.4 GHz band. Meanwhile, the ZigBee standard supports 868, 915 MHz and 2.4 GHz bands, and the 2.4 GHz ZigBee devices are deployed most widely in the world.

We focus on the coexistence of WiFi and ZigBee in the 2.4 GHz frequency band. The two kinds of networks have different channel specifications. WiFi has thirteen 20 MHz channels with 25 MHz channel spacing, while only three of them are nonoverlapping, according to the 802.11-2007 standard [18]. ZigBee has sixteen 2 MHz channels with 5 MHz channel spacing, numbering from 11 to 26, according to the 802.15.4-2006 standard [17]. Each WiFi channel overlaps with four ZigBee channels, as shown in Fig. 1.1. Due to the limited spectrum resources, the channel overlapping situation is unavoidable as each ZigBee channel would be covered by multiple WiFi channels.

1.2.1.2 The Physical Layer Specifications

Besides channel, the two kinds of devices adopt different physical layer specifications, as shown in Table 1.1. At first, they have different modulation types. ZigBee adopts OQPSK (Offset Quadrature Phase Shift Keying) while WiFi adopts OFDM (Orthogonal Frequency Division Multiplexing) with BPSK (Binary Phase Shift Keying), QPSK (Quadrature Phase Shift Keying), or QAM (Quadrature Amplitude Modulation) modulation; this difference makes the heterogeneous devices hard to

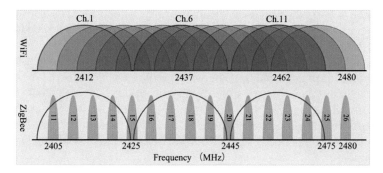

Fig. 1.1 Illustration of WiFi and ZigBee channels

Table 1.1 The physical layer parameters of ZigBee and WiFi

Technology	ZigBee	WiFi
IEEE standard	802.15.4	802.11
Modulation	OQPSK	OFDM+BPSK/QPSK/QAM
Transmission rate	250 kbps	> 2 Mbps
Transmission power	−3∼6 dBm	12∼20 dBm

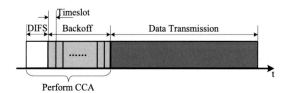

Fig. 1.2 An illustration of CSMA/CA

communicate with each other directly. In addition, they have asymmetry transmission power. Most ZigBee devices have the transmission power of about $0\,dBm$ to cut down energy consumption, while the WiFi devices have much higher transmission power with the purpose of large coverage; this difference leads to asymmetry cross-technology interference, which will be discussed later.

1.2.1.3 The MAC Layer Mechanism

In the MAC layer, both the WiFi and ZigBee networks adopt CSMA/CA (carrier sense multiple access with collision avoidance) mechanism to contend the channel. The detailed CSMA/CA mechanism is shown in Fig. 1.2. When a device begins to transmit a data packet, it first waits for DIFS (DCF Interframe Space) time; if the channel is idle during DIFS, the device then waits for a random duration which consists of multiple backoff timeslots to contend for the channel; the backoff timer is decreased by one when the channel is idle for a backoff slot, and is frozen when the channel is busy; the device can finally transmit a data packet if the backoff timer reaches zero. During DIFS or each backoff timeslot, the device should conduct CCA (Clear Channel Assessment) to determine whether the channel is idle. This process is quite simple: the channel is determined to be idle if the detected signal energy is below a predefined threshold β_E, otherwise it is busy.

The main difference here between WiFi and ZigBee is that, the WiFi DIFS is $28\,\mu s$ [19] while ZigBee DIFS is $320\,\mu s$ [21], meanwhile, WiFi backoff slot is 9 or $20\,\mu s$ while ZigBee backoff slot is $320\,\mu s$. This leads to extreme unfairness in the channel competition, as the WiFi device can always win the channel for transmission.

1.2.2 Cross-Technology Coexistence Problem

The two kinds of heterogeneous devices lead to severe cross-technology coexistence problems, such as the cross-technology interference (CTI) management and cross-technology communication (CTC), which dramatically affect the network performance as well as the performance of upper layer applications. Here we use a smart home scenario where the WiFi and ZigBee devices coexist to illustrate the two issues, as shown in Fig. 1.3.

The CTI issue exists as the devices access the wireless channel independently in a random way, they have a high probability to interfere with each other since the heterogeneous devices are not coordinated efficiently. As shown in Fig. 1.3, the

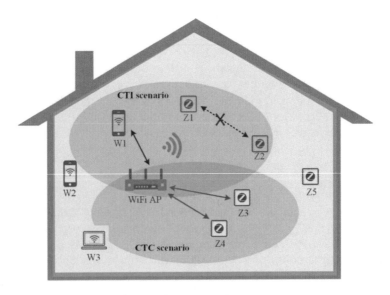

Fig. 1.3 The heterogeneous devices coexistence scenario

WiFi link $AP \longleftrightarrow W1$ and the ZigBee link $Z1 \longleftrightarrow Z2$ may perform their data transmissions concurrently, as the WiFi devices can not always sense the ZigBee transmission due to the low power. At this time, the ZigBee link is more vulnerable to be interfered, and it may also lead to failure of the WiFi transmission due to the reduced SNR (signal to noise ratio).

CTC is an emerging topic in recent years. Through proper physical layer design, CTC establishes direct communication between heterogeneous devices to improve the network performance and boost new applications [25]. For example, in Fig. 1.3, through CTC information exchange, WiFi AP can coordinate the heterogeneous devices $Z3$ and $Z4$ explicitly to avoid CTI and improve the network performance. In addition, since AP can exchange information with ZigBee devices directly through CTC, the two kinds of networks can share information with each other to provide context-aware services (e.g., $Z3$ can turn to the "away mode" when the WiFi AP discerns that the resident is not at home and notifies this information to $Z3$).

From the two scenarios we see that, signal identification is very important to solve both CTI and CTC issues better. Since the CTC mechanism is technology-specific, it is essential for a device to obtain the heterogeneous signal type and channel at first and then adopt a proper mechanism for CTC data transmission and reception. As to CTI, a device can avoid interference effectively when the heterogeneous signal type and channel is identified correctly, such as reserving channels for heterogeneous devices or adjusting the channel access operations.

1.3 Literature Review

This part reviews the works related to this book from three categories, including heterogeneous signal identification, cross-technology interference (CTI) management and cross-technology communication (CTC).

1.3.1 Heterogeneous Signal Identification

Besides helping solve CTI and CTC issues, signal identification has broader applications in wireless networks, such as dynamic spectrum access, spectrum interference monitoring, and opportunistic mesh networking [30]. Recent years have seen some advances in this field.

Some works extract the signal features for identification. SoNIC [13], ZiSense [47], and TIIM [14] exploit the time-domain features, such as RSSI (Received Signal Strength Indicator) and LQI (Link Quality Indicator), to classify WiFi signals on ZigBee devices. Airshark [31] utilizes RSSI as well as spectrum features to classify Bluetooth, ZigBee and microwave oven signals. The main problem of using time-domain signal features is that, the methods always require a group of data packets for signal identification, leading to a long time in signal collection.

Some works exploit deep learning for signal identification. Authors in [30] utilize CNN (Convolutional Neural Network) and deep RN (Residual Network) models to classify signals with different modulation types. Authors in [32] sense radar signals and estimate their spectrum occupancy through object detection on the spectrum images. These works show us a possibility of using deep learning to achieve effective signal identification. However, to the best of our knowledge, so far there is no signal identification method for the real WiFi and ZigBee signals collected through hardware testbed.

1.3.2 Cross-Technology Interference Management

The research on CTI management has a long history, and existing works mainly fall into two categories: interference resistance and interference avoidance.

Interference resistance mechanisms combat CTI through utilizing physical layer technologies. For example, BuzzBuzz [28] increases successful ZigBee transmissions under WiFi interference through designing ZigBee packets with more redundancy. ZIMO [40] utilizes the MIMO (Multiple-Input Multiple-Output) technology to transmit WiFi and Zigbee signals separately through different data streams. CrossZig [15] and PolarScout [33] allow ZigBee devices recover the collided packets according to the interference features through physical layer design. These schemes always require hardware modifications or even new transceiver design, which cannot be applied to current devices.

Interference avoidance has attracted much more research interest, through time-domain or frequency-domain mechanisms. The time-domain mechanisms intend to improve the network performance through modifying the MAC layer protocols, and can be further divided into two categories: uncoordinated and coordinated mechanisms. The uncoordinated mechanisms usually let a device identify the signal or interference and then make proper channel access decisions. With the signal identification processes described in Sect. 1.3.1, a device can apply a better coexistence strategy to improve the network performance, such as predicting white space prediction, exploiting concurrent transmissions, and channel switching [13, 14, 16, 29, 47]. The coordinated mechanisms avoid CTI through exchanging coordinated information among heterogeneous devices. WiCop[35], CBT[44] and Weeble[1] make ZigBee devices transmit special signals to improve their visibility to WiFi, so that WiFi devices can keep silence to avoid interference. Some methods utilize CTC technologies to convey the coordination information [6, 8, 26, 38, 42, 43, 46]. For example, ECC [42] makes a WiFi AP coordinate all the WiFi and ZigBee transmissions to avoid interference; BiCord [43] utilizes the coordination information to achieve efficient RF channel allocation. NetCTC [36] and CRF [37] leverages CTC to improve the network performance on unicast, multicast and flooding.

The frequency-domain mechanisms intend to avoid CTI through making heterogeneous devices working on frequency bands with no mutual interference. CoHop [39] designs channel hopping for ZigBee devices based on the calculated channel quality. G-Bee [2] makes a ZigBee device transmit data packet on the guard band of WiFi traffic working on non-overlapped channels. EmBee [5] makes a WiFi device reserve the channel for ZigBee transmission through designing null subcarriers on the ZigBee channel. These mechanisms require substantial modifications on either the MAC layer protocol or hardware design.

1.3.3 Cross-Technology Communication

Previous works on CTC design in the WiFi and ZigBee coexisting network can be divided into two categories: packet-level energy modulation and physical-layer CTC.

CTC through packet-level energy modulation has a long history, and the first work was proposed by Esense [3] which designs an "alphabet set" through a group of packet transmission duration. HoWiES [45] extends the basic idea of Esense for WiFi energy saving through using a low-power ZigBee radio to wake up the high-power WiFi interface. GSense [46] redesigns the WiFi preamble with a sequence of energy pulses, and uses the periods between pulses to transmit CTC information. FreeBee [24], C-Morse [41] and DCTC [23] utilize the WiFi beacon intervals or a set of predefined WiFi packet duration to convey CTC information. EMF [38] and WiZig [7] propose the upper layer design to improve the CTC performance. StripComm [48] introduces Manchester Coding to the CTC context to combat interference. This kind of mechanisms have low CTC data rate; meanwhile, they require time slot allocation for packet transmission, making them incompatible with WiFi or ZigBee MAC layer protocol design.

The physical-layer CTC was firstly proposed by WEBee [25] to make a commercial WiFi device elaborately construct the WiFi payload to transmit a ZigBee-compliant packet through signal emulation, which would then be detected by a ZigBee device directly. It can achieve high CTC data rate up to the ZigBee data rate. TwinBee [4] and LongBee [27] were further proposed to combat the high packet error rate of WEBee, so as to improve its reliability and transmission range. PAR [12] improves the CTC reliability through using a feedback channel. WIDE [11] achieves CTC through digital emulation to avoid signal distortion in the previous signal emulation works. The main problem of these mechanisms is that they can not work under the standard ZigBee and 20 MHz WiFi channels, as CTC can only be achieved when either the WiFi or ZigBee channel is changed to a non-standard one, to avoid the situation of the ZigBee channel overlapping with a WiFi pilot or null subcarrier.

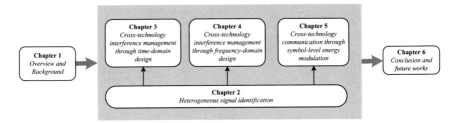

Fig. 1.4 The book structure

1.4   Book Structure

In this book, we design a series of mechanisms to combat the cross-technology coexistence problem in heterogeneous wireless networks. In particular, we introduce background of heterogeneous wireless technologies and also the cross-technology coexistence problem, then present a framework to describes the protocol design in this book. According to the framework, we first present a heterogeneous signal identification method based on both signal feature extraction and deep learning approaches; this is the basis for the following design. Then, we present two cross-technology interference management mechanisms in both the time-domain and frequency-domain, to improve the network performance. We also present a cross-technology communication mechanism through symbol-level energy modulation to boost new applications and support new protocol design. Finally, we discuss some possible research directions to further improve the efficiency of cross-technology coexistence.

The relationship among the six chapters is described as Fig. 1.4. More specifically, this chapter provides the background and literature review of cross-technology coexistence problem. Chapter 2 presents heterogeneous signal identification methods. Chapters 3–4 propose two protocols to achieve effective CTI management from time-domain and frequency-domain system design. Chapter 5 proposes a CTC mechanism to further improve the efficiency of cross-technology coexistence. Lastly, Chap. 6 concludes this book and points out possible future research directions.

References

1. Bozidar, R., Chandra, R., Gunawardena, D.: Weeble: enabling low-power nodes to coexist with high-power nodes in white space networks. In: Proceedings of the ACM CoNEXT (2012)
2. Chae, Y., Wang, S., Kim, S.M.: Exploiting WiFi guard band for safeguarded ZigBee. In: Proceedings of the ACM SenSys (2018)
3. Chebrolu, K., Dhekne, A.: Esense: communication through energy sensing. In: Proceedings of the ACM MobiCom (2009)

4. Chen, Y., Li, Z., He, T.: TwinBee: reliable physical-layer cross-technology communication with symbol-level coding. In: Proceedings of the IEEE INFOCOM (2018)
5. Chen, R., Gao, W.: Enabling cross-technology coexistence for extremely weak wireless devices. In: Proceedings of the IEEE INFOCOM (2019)
6. Chen, W., Yin, Z., He, T.: Global cooperation for heterogeneous networks. In: Proceedings of the IEEE INFOCOM (2020)
7. Chi, Z., Huang, Z., Yao, Y., Xie, T., Sun, H., Zhu, T.: EMF: embedding multiple flows of information in existing traffic for concurrent communication among heterogeneous IoT devices. In: Proceedings of the IEEE INFOCOM (2017)
8. Chi, Z., Li, Y., Huang, Z., Sun, H., Zhu, T.: Simultaneous bi-directional communications and data forwarding using a single ZigBee data stream. In: Proceedings of the IEEE INFOCOM (2019)
9. Cisco: Cisco annual internet report (2018–2023) white paper (2020)
10. Guo, X., Zheng, X., He, Y.: WiZig: cross-technology energy communication over a noisy channel. In: Proceedings of the IEEE INFOCOM (2017)
11. Guo, X., He, Y., Zhang, J., Jiang, H.: WIDE: physical-level CTC via digital emulation. In: Proceedings of the ACM/IEEE IPSN (2019)
12. He, H., Su, J., Chen, Y., Li, Z., Li, L.: Reliable cross-Technology communication with physical-layer acknowledgement. IEEE Trans. Commun. **68**(8) (2020)
13. Hermans, F., Rensfelt, O., Voigt, T., Ngai, E., Nordén, L., Gunningberg, P.: SoNIC: classifying interference in 802.15.4 sensor networks. In: Proceedings of the IEEE IPSN (2013)
14. Hithnawi, A., Shafagh, H., Duquennoy, S.: TIIM: technology-independent interference mitigation for low-power wireless networks. In: Proceedings of the IEEE IPSN (2015)
15. Hithnawi, A., Li, S., Shafagh, H., Gross, J., Duquennoy, S.: CrossZig: combating cross-technology interference in low-power wireless networks. In: Proceedings of the ACM/IEEE IPSN (2016)
16. Huang, J., Xing, G., Zhou, G., Zhou, R.: Beyond co-existence: exploiting WiFi white space for Zigbee performance assurance. In: Proceedings of the IEEE ICNP (2010)
17. IEEE Computer Society. 802.15.4: Wireless medium access control (MAC) and physical layer (PHY) specifications for low-rate wireless personal area networks (WPANs) (2006)
18. IEEE Computer Society. 802.11: Wireless LAN medium access control (MAC) and physical layer (PHY) specifications (2007)
19. IEEE Computer Society. 802.11: Wireless LAN medium access control (MAC) and physical layer (PHY) specifications – Amendment 5: enhancements for higher throughput (2009)
20. IEEE Computer Society. 802.11: Wireless LAN medium access control (MAC) and physical layer (PHY) specifications – Amendment 4: enhancements for very high throughput for operation in bands below 6 GHz (2013)
21. IEEE Computer Society. 802.15.4: IEEE Standard for low-rate wireless networks (2016)
22. IEEE Computer Society. 802.11: Wireless LAN medium access control (MAC) and physical layer (PHY) specifications – Amendment 1: enhancements for high-efficiency WLAN (2021)
23. Jiang, W., Yin, Z., Kim, S.M., He, T.: Transparent cross-technology communication over data traffic. In: Proceedings of the IEEE INFOCOM (2017)
24. Kim, S.M., He, T.: FreeBee: cross-technology communication via free side-channel. In: Proceedings of the ACM MobiCom (2015)
25. Li, Z., He, T.: WEBee: physical-layer cross-technology communication via emulation. In: Proceedings of the ACM MobiCom (2017)
26. Li, Y., Chi, Z., Liu, X., Zhu, T.: Chiron: concurrent high throughput communication for iot devices. In: Proceedings of the ACM MobiSys (2018)
27. Li, Z., He, T.: LongBee: enabling long-range cross-technology communication. In: Proceedings of the IEEE INFOCOM (2018)
28. Liang, C.-J.M., Priyantha, N.B., Liu, J., Terzis, A.: Surviving wi-fi interference in low power ZigBee networks. In: Proceedings of the ACM SenSys (2010)
29. Meng, J., He, Y., Zheng, X., Fang, D., Xu, D., Xing, T., Chen, X.: Smoggy-Link: fingerprinting interference for predictable wireless concurrency. In: Proceedings of the IEEE ICNP (2016)

30. O'Shea, T.J., Roy, T., Clancy, T.C.: Over-the-air deep learning based radio signal classification. IEEE J. Sel. Top. Signal Process. **12**, 168–179 (2018)
31. Rayanchu, S., Patro, A., Banerjee, S.: Airshark:detecting non-WiFi RF devices using commodity WiFi hardware. In: Proceedings of the ACM IMC (2011)
32. Sarkar, S., Buddhikot, M., Baset, A., Kasera, S.K.: Deepradar: a deep-learning-based environmental sensing capability sensor design for cbrs. In: Proceedings of the ACM MobiCom (2021)
33. Shao, C., Park, H., Roh, H., Lee, W., Kim, H.: PolarScout: Wi-Fi interference-resilient ZigBee communication via shell-shaping. IEEE/ACM Trans. Netw. **28**(4), 1587–1600 (2020)
34. Verified Market Research: Global zigbee market size by standards, by application, geographic scope and forecast (2020)
35. Wang, Y., Wang, Q., Zeng, Z., Zheng, G., Zheng, R.: Wicop: engineering wifi temporal white-spaces for safe operations of wireless body area networks in medical applications. In: Proceedings of the IEEE RTSS (2011)
36. Wang, S., Yin, Z., Li, Z., He, T.: Networking support for physical-layer cross-technology communication. In: Proceedings of the IEEE ICNP (2018)
37. Wang, W., Liu, X., Yao, Y., Pan, Y., Chi, Z., Zhu, T.: CRF: coexistent routing and flooding using WiFi packets in heterogeneous IoT networks. In: Proceedings of the IEEE INFOCOM (2019)
38. Wang, W., Xie, T., Liu, X., Zhu, T.: ECT: exploiting cross-technology transmission for reducing packet delivery delay in IoT networks. ACM Trans. Sensor Netw. **15**(2) (2019)
39. Wang, Y., Zheng, X., Liu, L., Ma, H.: CoHop: quantitative correlation-based channel hopping for low-power wireless networks. ACM Trans. Sensor Netw. **17**(2), 1–29 (2021)
40. Yan, Y., Yang, P., Li, X., Tao, Y., Zhang, L., You, L.: ZIMO: building cross-technology MIMO to harmonize zigbee smog with WiFi flash without intervention. In: Proceedings of the ACM MobiCom (2013)
41. Yin, Z., Jiang, W., Kim, S.M., He, T.: C-Morse: cross-technology communication with transparent Morse coding. In: Proceedings of the IEEE INFOCOM (2017)
42. Yin, Z., Li, Z., Kim, S.M., He, T.: Explicit channel coordination via cross-technology communication. In: Proceedings of the ACM MobiSys (2018)
43. Yu, Z., Li, P., Boano, C.A., He, Y., Jin, M., Guo, X., Zheng, X.: BiCord: bidirectional coordination among coexisting wireless devices. In: Proceedings of the IEEE ICDCS (2021)
44. Zhang, X., Kang, G.S.: Enabling coexistence of heterogeneous wireless systems: case for ZigBee and WiFi. In: Proceedings of the ACM MobiHoc (2011)
45. Zhang, Y., Li, Q.: HoWiES: a holistic approach to ZigBee assisted WiFi energy savings in mobile devices. In: Proceedings of the IEEE INFOCOM (2013)
46. Zhang, X., Shin, K.G.: Gap sense: lightweight coordination of heterogeneous wireless devices. In: Proceedings of the IEEE INFOCOM (2013)
47. Zheng, X., Cao, Z., Wang, J., He, Y., Liu, Y.: ZiSense: towards interference resilient duty cycling in wireless sensor networks. In: Proceedings of the ACM SenSys (2014)
48. Zheng, X., He, Y., Guo, X.: StripComm: interference-resilient cross-technology communication in coexisting environments. In: Proceedings of the IEEE INFOCOM (2018)

Chapter 2
Heterogeneous Signal Identification

Abstract In this chapter, we will present the motivation of signal identification for cross-technology coexistence, then elaborate our system design which contains two methods for signal identification. We will finally give performance evaluation for the two methods based on signals collected from hardware testbed.

Keywords Signal identification · FFT · CNN

2.1 Introduction

The WiFi and ZigBee devices operating on the same 2.4 GHz ISM band will lead to severe cross-technology coexistence problem, degrading the network performance. We consider the sensing of signal type as well as the signal channel can assist wireless devices with more efficient transmission strategy. For example, when a high power WiFi device senses a ZigBee signal, it can void its interference to these devices through adjusting its transmission power or channel.

We have seen some previous works on signal identification for network performance improvement [1–11]. Authors in [8] identify the signal type of WiFi, ZigBee, BlueTooth, and microwave oven signals mainly through analyzing the time-domain features, like the on-air time and inter-packet duration; however, these features highly depend on the data traffic and device movements, they cannot be used in complex mobile scenarios, and also can not be used to identify the channel of signals. Some works exploit deep learning technologies for signal identification. Authors in [7] identify the modulation types of signals through feeding the raw signal samples to the CNN (convolutional neural network) and deep RN (residual network) models; authors in [9] sense radar signals and estimate their spectrum occupancy through object detection on the spectrum. So far there is no effective signal identification for WiFi and ZigBee signals. This chapter proposes ZShark, a fast signal identification method to achieve this goal. We leave the cross-technology coexistence design based on signal identification in the following chapters.

2.2 Motivation

The opportunity for heterogeneous signal identification is from the key insight that, frequency-domain features are much more stable during a packet transmission, as they are little affected by the data traffic and device movements. We conduct experiments to investigate the frequency spectrum characteristics, and the results indicate that, even when the SINR (signal to interference and noise ratio) is below zero and the ZigBee signal duration is only 3.2 μs (corresponding to 64 sample length), the frequency spectrum can still exhibit distinguishable features. As shown in Fig. 2.1a, the ZigBee signal is submerged in the WiFi signal, but the ZigBee and WiFi collided signal has much different frequency-domain features compared to the pure WiFi signal, as shown in Fig. 2.1b, c. This difference inspires us to further study the performance of heterogeneous signal identification through using the spectrum features. One may ask why the frequency-domain feature is identifiable but time-domain feature is not. The reason is that, ZigBee signal occupies 2 MHz frequency band but WiFi occupies 20 MHz; when a WiFi device receives and samples a ZigBee signal, the ZigBee signal power concentrated in 2 MHz will be averaged in the whole 20 MHz band, leading to much lower energy in the time-domain.

The main challenge on quantitative analysis of signal identification is to construct dataset correctly based on signal collected from real networks, since the ZigBee

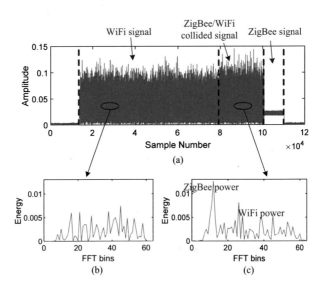

Fig. 2.1 An example of frequency spectrum features with short sample length. In the collected signal, a ZigBee packet is overlapped with a WiFi packet in some parts; since the ZigBee signal has low amplitude comparing with the WiFi signal, it is hard to be identified through time-domain features. However, in the frequency domain, the ZigBee signal exhibits much higher energy than WiFi. (**a**) Amplitude of the overlapped signal. (**b**) Frequency spectrum of WiFi signal. (**c**) Frequency spectrum of interfered ZigBee signal

signal is hard to be extracted through intuitive ways such as energy change, as shown in Fig. 2.1a. In this paper, we propose ZShark to construct the dataset with the aid of physical layer technologies, based on which we propose signal identification mechanisms and evaluate their performance.

2.3   System Design

In this section, we give the system design of ZShark for fast signal identification at WiFi devices. We first describe the process of constructing the dataset, then propose two methods, Zshark-FFT and Zshark-CNN, for signal identification.

2.3.1   Dataset Construction

We first show the data collection process based on hardware testbed, then illustrate how the WiFi and ZigBee signals are extracted to construct the dataset.

2.3.1.1   Data Collection

We first set up a testbed to collect heterogeneous signals at WiFi devices, as shown in Fig. 2.2. Specifically, we make a USRP (Universal Software Radio Peripheral) N210 be a WiFi signal collector (called USRP N210-1). It collects signals transmitted from another USRP N210 (called USRP N210-2) and a commercial ZigBee platform named TelosB. We set the sample rate of the two USRP devices be 20 MHz. The WiFi channel has the central frequency of 2.412 GHz, which overlaps with four ZigBee channels varying from 11 (the central frequency is 2.405 GHz) to 14 (the central frequency is 2.420 GHz). We adjust the distance between TelosB and USRP

Fig. 2.2 The data collection testbed

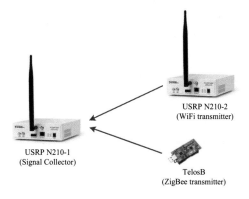

N210-1, as well as the transmission power of TelosB and USRP N210-2, to generate the desired SINR and receiving power at the signal collector (USRP N210-1).

A wireless signal is typically described as a stream of complex samples, also called I/Q (In-phase/Quadrature) samples. The bit stream of a WiFi/ZigBee packet will be modulated into a series of complex symbols, and transmitted after DAC (Digital to Analog Convertor) and modulating a carrier signal with high frequency f_c. Here $f_c = 2.412\,\text{GHz}$ for WiFi and $f_c = 2.405/2.410/2.415/2.420\,\text{GHz}$ for ZigBee. At the receiver side, a device first down-converts the signal with frequency f_c to obtain the baseband signal, then passes the signal through an ADC (Analog to Digital Convertor) module to obtain the complex I/Q samples for the following data demodulation. In this testbed, we set $f_c = 2.412\,\text{GHz}$ for USRP N210-1 and N210-2. The data collected by USRP N210-1 are the raw I/Q samples after ADC.

2.3.1.2   Signal Extraction

The raw I/Q samples collected by USRP N210 may contain the ZigBee signal, WiFi signal and background noise. We first extract the WiFi and ZigBee signals from the raw I/Q samples, then partition and label the samples to construct the dataset.

(i) ZigBee Signal Extraction

Suppose x_i is the complex number that represents the i-th transmitted ZigBee signal at the baseband, the corresponding received sample at WiFi can be denoted as $s_i = H \cdot x_i e^{j2\pi i \Delta f_z T_s}$, where H is the channel attenuation factor, T_s is the sampling interval and here $T_s = \frac{1}{20\,\text{MHz}} = 0.05\,\mu s$, Δf_Z is the CFO (Central Frequency Offset) between the ZigBee transmitter (TelosB) and WiFi receiver (USRP N210-1), and it may roughly be $-7, -2, 3$ or $8\,\text{MHz}$, according to the central frequencies of the transmitter and receiver.

The ZigBee preamble contains eight repetitive symbols '0000', each symbol lasts for $16\,\mu s$, corresponding to 320 I/Q samples when collected at WiFi devices, whose sample rate is 20 MHz. That means, the first four symbols (corresponding to $320 \times 4 = 1280$ samples) are exactly the same as the second set of four symbols, that is, $x_{i+1280} = x_i$. We utilize this feature for ZigBee signal extraction.

When we conduct cross correlation of four adjacent symbols '0000' (correspond to $320 \times 4 = 1280$ samples) with the following four ones, the correlation result at m-th sample is calculated as:

$$|C_m^Z| = \sum_{i=0}^{1279} \left| s_{m+i} \cdot s_{m+i+1280}^* \right|, \tag{2.1}$$

where s_m^* is the complex conjugate of s_m.

Fig. 2.3 The correlation results calculated through Eq. (2.2)

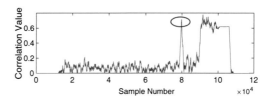

If m' is the first sample of ZigBee preamble, the correlation result will have a peak value here due to the high correlation characteristic, that is:

$$|C_{m'}^Z| = \sum_{i=0}^{1279} \left| H x_{m'+i} e^{j2\pi \Delta f_z (m'+i)T_s} \cdot H x_{m'+i+1280}^* e^{-j2\pi \Delta f_z (m'+i+1280)T_s} \right|$$

$$= \sum_{i=0}^{1279} \left| H^2 x_{m'+i} \cdot x_{m'+i}^* \cdot e^{-j2\pi \Delta f_z 1280 T_s} \right|$$

$$= \sum_{i=0}^{1279} \left| H^2 x_{m'+i} \right|^2 ,$$

$$(2.2)$$

while in other positions, the value $|C_m^Z|$ will be much lower than $|C_{m'}^Z|$, as shown in Fig. 2.3, which indicates the correlation results at each position calculated by Eq. (2.2) for the original signal in Fig. 2.1a. As to other signals without the ZigBee preamble, the correlation results will always be very low. With this feature, it is easy to determine the beginning of a ZigBee packet and extract it from the received signal.

(ii) WiFi Signal Extraction

The WiFi preamble contains 10 repetitive STS (Short Training Symbols) for crude CFO estimation, each STS lasts for 0.8 μs, corresponding to 16 samples under the 20 MHz sample rate. The WiFi signal extraction process is similar to ZigBee. Since the first five STS symbols (correspond to $16 \times 5 = 80$ samples) are exactly the same as the following five ones, here $x_{i+80} = x_i$.

When we conduct cross correlation of 80 samples with the following 80 ones, the correlation result at m-th sample is calculated as:

$$|C_m^W| = \sum_{i=0}^{79} \left| s_{m+i} \cdot s_{m+i+80}^* \right| .$$

$$(2.3)$$

The highest correlation result will occur at the first sample of WiFi preamble, that is $C_{m'}^W = \sum_{i=1}^{80} H^2 \left| x_{m'+i} \right|^2$. We see a peak value at the beginning of the WiFi packet in Fig. 2.4, which exhibits the correlation results at each position calculated

Fig. 2.4 The correlation
results calculated through
Eq. (2.3)

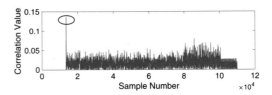

by Eq. (2.3) for the original signal in Fig. 2.1a. In other situations when the signals
have no repetitive STS, the correlation results will always be very low. Thus, it is
easy to extract WiFi signal from the collected signal with this feature.

2.3.1.3 Dataset Description

We can obtain the beginning positions of both WiFi and ZigBee signals in the
collected samples through the signal extraction process. Since the signal duration
for each ZigBee and WiFi packet is known in the experiment, we can easily extract
the two kinds of signals successfully, even under low SNR/SINR situations. We then
partition each signal into equal parts with the sample length L_s. Considering the
requirements for time-domain design in Chap. 3, we let $L_s = 64$, which corresponds
to $3.2\mu s$ duration and it is much shorter than a backoff slot ($9\,\mu s$).

2.3.2 Signal Identification Design

This part proposes ZShark for heterogeneous signal identification. It contains
two methods, ZShark-FFT which uses the extracted frequency spectrum features
based on FFT (Fast Fourier Transform), and ZShark-CNN which uses the CNN
(convolutional neural network) model.

2.3.2.1 Signal Identification Through FFT Features

As shown in Fig. 2.1, the WiFi and ZigBee signals have distinguishable frequency
spectrum features, even under low SINR situations. We see that on the 20 MHz WiFi
channel, the ZigBee signals have much higher energy on the 2 MHz overlapped
ZigBee channel than that on the non-overlapped channel, while the WiFi signals
have stable energy in the whole channel. Thus, we choose the metric, *Ratio of
Energy on overlapped and non-overlapped channels (RoE)*, as the key feature to
differentiate the two kinds of signals.

For a sequence of samples $S = \{s_1, s_2, \ldots, s_N\}$ in the dataset, we first conduct
$N-$point FFT to obtain its frequency spectrum, that is, $F_k = \sum_{n=1}^{N} s_n e^{-j2\pi k \frac{n}{N}}, k = 1, \ldots, N$. Then, the signal energy on the ZigBee overlapped channel is $EO_{ch_l} =$

$\sum_{<i>} F_i^2$, where $l = 1, \ldots, 4$ and i refers to the frequency point overlapped with the l-th ZigBee channel.[1] The signal energy on the ZigBee non-overlapped channel is $EN = \sum_{<k>} F_k^2$, here k refers to the frequency point which is not overlapped with the ZigBee channel. The parameters of l, i, and k are easily to be obtained with the WiFi and ZigBee channels.

Then RoE of this sample sequence is calculated as:

$$RoE = \frac{max\{EO_{ch_l}\}}{EN}, l \in [1, 4]. \tag{2.4}$$

The WiFi signal and ZigBee signals operating on different channels will lead to very different RoE values. We finally use logistic regression to complete signal identification according to the calculated features.

2.3.2.2 Signal Identification Through Deep Learning

Considering that CNN (convolutional neural network) has the high capability of extracting hidden features, and previous works have proven its feasibility on classifying signals with different modulations [7], here we further propose ZShark-CNN on the raw signals for identification.

The convolutional neural network (CNN) includes a set of convolutional layers and several fully-connected layers. The convolutional layer is used for feature extraction and the latter is used for classification. CNN has wide applications in natural language processing, computer vision, speed recognition, and the wireless networking area. We start from the CNN model proposed in [7], as it is more related to this context. Meanwhile, as the spectrum features we intend to use here is more apparent than the modulation features used in [7], we expect a simpler network with lower computation complexity and less energy consumption. Thus, we will simplify the model in [7] and evaluate the performance of identifying heterogeneous wireless signals.

The CNN architecture is shown in Fig. 2.5, while Table 2.1 illustrates a set of CNN network configurations investigated in this work, from Net A which is relatively complex to Net E which is very simple. The input is a sequence of complex samples with length $L_s = 64$. They are transformed to a 64×2 matrix containing the real and imaginary parts of the 64 samples. The amplitude of the input samples is normalized to mitigate the impact of signal strength. The input samples are fed into the convolutional layers (Conv2-n in Table 2.1, where n indicates the number of filters) with filter size $(1, 3)$. We use the activation function of rectified linear (ReLU) for each layer to perform non-linear transformation. The multi-dimensional data output from the last convolution layer is reshaped as a one-dimensional vector after the flatten layer, and then sent into the fully-connected layer FC-m in Table 2.1,

---

[1] Each WiFi channel overlaps with four ZigBee channels in the same pattern, as shown in Fig. 1.1.

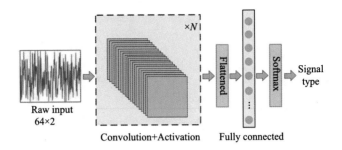

Fig. 2.5 The CNN architecture

Table 2.1 CNN network configurations

Net A	Net B	Net C	Net D	Net E
Input (64*2)				
Conv2-64	Conv2-16	Conv2-4	Conv2-4	Conv2-2
Conv2-128	Conv2-32	Conv2-8		
FC-256	FC-128	FC-32	FC-16	FC-8

where m indicates the size of output after this layer. Softmax is finally used for classification.

2.4 Performance Evaluation

2.4.1 Experimental Settings

The first important step to evaluating the performance is to construct proper dataset, through the testbed shown in Fig. 2.2. We let USRPs operate on the 11th WiFi channel, and let the TelosB platform operate on each of the four overlapped ZigBee channels numbering from 21 to 24, as shown in Fig. 1.1. From the aforementioned analysis, we see that the ZigBee signal identification accuracy can be highly affected by the background interference P_{BG} and the ZigBee signal power P_r^Z. Thus, we collect a set of signals under different P_{BG} and SINR environments. As shown in Fig. 2.2, P_{BG} includes the receiving WiFi signal power at USRP N210-1 transmitted from USRP N210-2 (denoted by P_r^W) and the background noise (denoted by P_N). P_N is tested to be -95 dB in the experiment. The SINR of the ZigBee signal is calculated as $SINR_Z = \frac{P_r^Z}{P_{BG}}$. Thus, under a P_{BG}, we generate a specific $SINR_Z$ through adjusting TelosB transmission power and the distance between TelosB and USRP N210-1. For each situation of $SINR_Z$ and P_{BG}, we first collect the original signal at USRP N210-1, and utilize the signal extraction process described in Sect. 2.3.1.2 to obtain the pure WiFi signal and the ZigBee/WiFi collided signal. We then divide each kind of signal into sequences with $L_s = 64$ sample length, and label them for the signal identification models. Specifically, the WiFi signal is labeled as WiFi, and the ZigBee/WiFi collided signal is labeled as ZigBee with the corresponding channels. We finally construct a dataset with 36,000 WiFi sample sequences and 36,000 ZigBee sample sequences, corresponding to $72{,}000 \times L_s$ data samples in total. Among the dataset, 80% are training data, 20% are testing data.

We use two metrics, the false positive (FP) and false negative (FN) error rates, to measure the performance of ZigBee signal identification. The FP error means that a ZigBee signal on a specific channel is identified although it is not in the received signal on that channel, and the FN error means a ZigBee signal on a specific channel is not identified but it is in the received signal on that channel.

2.4.2 Experimental Results

We use the constructed dataset to evaluate the performance of both ZShark-FFT and ZShark-CNN.

(1) Performance of ZShark-FFT
The FP and FN error rates of ZigBee signal identification under different settings are shown in shows the false positive (FP) and false negative (FN). The notation '−−' indicates that setting an not be obtained accurately in the experiment. For example, when $P_{BG} = -91$ dB, the situation of $SINR_Z = 0$ dB is difficult to be obtained, as the ZigBee receiving power is drowned in the background noise and hard to be calculated. Therefore, we do not provide the corresponding results, but that doesn't mean that ZShark loses its function completely.

From Table 2.2 we see that, under $P_{BG} = -91$ dB and $SINR_Z = 2$ dB, which means there is no WiFi interference and the ZigBee receiving power is 2 dB higher than the background noise, the ZigBee signal can be identified with

Table 2.2 The FP/FN error rates of ZigBee signal identification under different settings when $L_s = 64$ samples

P_{BG}	$SINR_z$							
	4 dB	2 dB	0 dB	−2 dB	−4 dB	−6 dB	−8 dB	−10 dB
−75 dB	0.000/ 0.011	0.000/ 0.012	0.000/ 0.021	0.000/ 0.022	0.006/ 0.032	0.010/ 0.056	0.025/ 0.182	0.026/ 0.203
−79 dB	0.000/ 0.012	0.000/ 0.011	0.000/ 0.022	0.000/ 0.027	0.006/ 0.043	0.011/ 0.076	0.024/ 0.150	0.036/ 0.253
−83 dB	0.000/ 0.010	0.000/ 0.012	0.000/ 0.025	0.000/ 0.041	0.006/ 0.050	0.010/ 0.113	−	−
−87 dB	0.000/ 0.011	0.000/ 0.012	0.000/ 0.033	0.000/ 0.055	−	−	−	−
−91 dB	0.000/ 0.010	0.000/ 0.012	−	−	−	−	−	−

the probability of 98.8%. It can also be identified under low SINR environments when P_{BG} increases. For example, when P_{BG} is increased to -75 dB, the FN error rate is only 2.2% when $SINR_Z = -2$ dB, while the error rate increases dramatically with the decrease of $SINR_Z$ (the FN error rate is as high as 20.3% when $SINR_Z = -10$ dB). We also see that the FP error rate is always very low under each setting, which means nearly no WiFi signal is incorrectly identified as ZigBee.

We note that the ZigBee signal identification results include the signal type and channel identification, and the FP/FN error rates shown in Table 2.2 are the averaged values of the error rates calculated for the four ZigBee channels overlapped with the WiFi channel, as shown in Fig. 1.1. For example, when $P_{BG} = -75$ dB and $SINR_Z = -6$ dB, the FP error rate of 1% means the ZigBee signal on each of the four overlapped channels can be identified with the averaged probability of 99%, as the ZigBee signals on different channels also exhibit very different frequency features.

(2) Performance of ZShark-CNN
We further use the constructed dataset to evaluate the performance of ZShark-CNN. The FP and FN error rates of ZigBee signal identification using Net E, the simplest model in Table 2.1, are shown in Table 2.3. We see much better performance compared with Table 2.2. For example, when $P_{BG} = -75$ dB and $SINR_Z = -10$ dB, the FN error rate of Zhark-CNN using Net E is only 2.0%, while that of ZShark-FFT is as high as 20.3%. We also conduct experiments of the same settings under different environments, and the difference in experimental results is very small, which shows the robustness of this model. The reason is that the feature learned by the CNN model is the frequency spectrum, which is independent with environments. Thus, the pre-trained model based on the collected signal is also suitable to other environments and is expected to achieve high identification accuracy.

Table 2.3 The FP/FN error rates of ZigBee signal identification under different settings using Net E

P_{BG}	$SINR_z$							
	4 dB	2 dB	0 dB	-2 dB	-4 dB	-6 dB	-8 dB	-10 dB
-75 dB	0.000/ 0.000	0.000/ 0.000	0.005/ 0.009	0.008/ 0.010	0.009/ 0.011	0.011/ 0.019	0.015/ 0.019	0.016/ 0.020
-79 dB	0.000/ 0.000	0.000/ 0.000	0.008/ 0.017	0.012/ 0.017	0.012/ 0.017	0.013/ 0.020	0.015/ 0.022	0.020/ 0.023
-83 dB	0.000/ 0.000	0.000/ 0.000	0.010/ 0.016	0.011/ 0.017	0.015/ 0.020	0.016/ 0.020	–	–
-87 dB	0.000/ 0.000	0.000/ 0.000	0.011/ 0.019	0.015/ 0.020	–	–	–	–
-91 dB	0.000/ 0.000	0.000/ 0.000	–	–	–	–	–	–

Table 2.4 FLOPs under net
A to net E

Net A	Net B	Net C	Net D	Net E
3.2 M	207 K	14 K	1.6 K	0.8 K

We further use the metric of FLOPs (floating point operations) to measure the computational complexity of the CNN models from Net A to Net E in Table 2.1. The results in Table 2.4 shows that computational complexity decreases dramatically when the CNN model becomes simpler, and the FLOPs of Net E is even below 1 K. We note that the FLOPs of Net E is nearly equivalent to that of ZShark-FFT under the same sample length ($L_s = 64$), which is 0.768 K. Considering that current CPU is always at GFLOPS (Gigabit floating point operations per second) level and even higher, the signal identification process through Net E is expected to be below $1\mu s$.

2.5 Summary

In this chapter, we first present the motivation of signal identification for cross-technology coexistence, then propose the system design which contains ZShark-FFT and ZShark-CNN for fast signal identification. We also collect signals from hardware testbed to evaluate the performance of these two methods.

References

1. Chae, Y., Wang, S., Kim, S.M.: Exploiting WiFi guard band for safeguarded ZigBee. In: Proceedings of the ACM SenSys (2018)
2. Chen, R., Gao, W.: Enabling cross-technology coexistence for extremely weak wireless devices. In: Proceedings of the IEEE INFOCOM (2019)
3. Hermans, F., Rensfelt, O., Voigt, T., Ngai, E., Nordén, L., Gunningberg, P.: SoNIC: classifying interference in 802.15.4 sensor networks. In: Proceedings of the IEEE IPSN (2013)
4. Hithnawi, A., Shafagh, H., Duquennoy, S.: TIIM: technology-independent interference mitigation for low-power wireless networks. In: Procedings of the IEEE IPSN (2015)
5. Huang, J., Xing, G., Zhou, G., Zhou, R.: Beyond co-existence: exploiting WiFi white space for Zigbee performance assurance. In: Proceedings of the IEEE ICNP (2010)
6. Meng, J., He, Y., Zheng, X., Fang, D., Xu, D., Xing, T., Chen, X.: Smoggy-Link: fingerprinting interference for predictable wireless concurrency. In: Proceedings of the IEEE ICNP (2016)
7. O'Shea, T.J., Roy, T., Clancy, T.C.: Over-the-air deep learning based radio signal classification. IEEE J. Sel. Top. Signal Process. **12**, 168–179 (2018)
8. Rayanchu, S., Patro, A., Banerjee, S.: Airshark: detecting non-wifi RF devices using commodity wifi hardware. In: Proceedings of the ACM IMC (2011)
9. Sarkar, S., Buddhikot, M., Baset, A., Kasera, S.K.: Deepradar: a deep-learning-based environmental sensing capability sensor design for cbrs. In: Proceedings of the ACM MobiCom (2021)
10. Wang, Y., Zheng, X., Liu, L., Ma, H.: CoHop: quantitative correlation-based channel hopping for low-power wireless networks. ACM Trans. Sensor Netw. **17**(2), 1–29 (2021)
11. Zheng, X., Cao, Z., Wang, J., He, Y., Liu, Y.: ZiSense: towards interference resilient duty cycling in wireless sensor networks. In: Proceedings of the ACM SenSys (2014)

Chapter 3
Cross-Technology Interference Management Through Time-Domain Design

Abstract In this chapter, we will introduce the problem of cross-technology interference problem, and also the motivation of time-domain protocol design to combat this problem. We will then design E-CCA to improve the network performance with quantitative theoretical analysis, and finally evaluate its performance through simulations.

Keywords Cross-technology interference management · The MAC layer · Clear channel assessment

3.1 Introduction

WiFi and ZigBee devices typically utilize the CSMA/CA (Carrier Sense Multiple Access with Collision Avoidance) mechanism to avoid interference through time-domain design. Before transmitting, the device listens to the channel and performs CCA (Clear Channel Assessment) process to determine the channel availability: if the detected energy is over a predefined CCA threshold, the channel is busy; otherwise, it is idle and the device can proceed with the transmission. Due to heterogeneous application requirements, the two kinds of devices transmit signals at different power levels. The ZigBee devices always transmit signals at less than $1mW$ for energy reservation [9], while WiFi devices can transmit signals up to $100mW$ for larger coverage [2]. Since the CCA threshold is fixed in each device, this power asymmetry further deteriorates the CTI problem, as ZigBee devices are always prohibited from transmission by WiFi.

Researchers have paid much attention to this problem, mainly through coexistence protocol design in the MAC layer, and solutions can be categorized into two groups: coordinated and uncoordinated mechanisms. Coordinated mechanisms alleviate CTI through exchanging coordinated information among heterogeneous devices, such that devices can make proper decisions to avoid interference [1, 4, 5, 12, 15–18]. These mechanisms can achieve much higher performance but require substantial modifications on the standard MAC layer, as new packets should be designed for information exchange. Uncoordinated mechanisms let a device identify

J. Yao, K. Wu, *Cross-Technology Coexistence Design for Wireless Networks*,
SpringerBriefs in Computer Science, https://doi.org/10.1007/978-981-99-1670-2_3

the heterogeneous signal type at first and then make proper channel access decisions [3, 7, 8, 10, 13, 19]. The signal identification can be achieved either through utilizing some key features in heterogeneous signals, such as those extracted from RSSI (Received Signal Strength Indicator), LQI (Link Quality Indicator), and interference pattern [13], or through physical layer technologies [3]. These mechanisms can be achieved with existing packets, but are also incompatible with the standard CSMA/CA. Both kinds of mechanisms are very hard to be deployed in current devices.

In this chapter, we present E-CCA, an uncoordinated time-domain mechanism which enhances the WiFi CCA process based on fast signal identification proposed in Sect. 2, so as to mitigate CTI in heterogeneous wireless networks, thus to improve the overall network performance. We evaluate the performance of E-CCA through both theoretical analysis and simulations on NS-3. The results indicate that the performance of ZigBee can be largely increased, with little sacrifice to WiFi.

3.2   Motivation

In this section, we present the cross-technology interference (CTI) issue, and also describe the opportunity of time-domain design to mitigate the interference.

Both WiFi and ZigBee devices adopt CSMA/CA to determine the channel access. However, the WiFi signal power is much higher than ZigBee while the WiFi channel bandwidth is much broader than ZigBee. Since WiFi devices do not differentiate the arriving ZigBee and WiFi signals, they use the same CCA threshold for the two kinds of signal, leading to asymmetry in CTI. As shown in Fig. 3.1, in the case that the WiFi and ZigBee transmitters, W_T and Z_{T1}, are within the carrier sense range of each other, current CSMA/CA mechanism can largely avoid CTI as both transmitters determine the channel to be busy; in the case that both the transmitters, W_T and Z_{T3}, are out of the carrier sense range of each other, CTI does not occur

Fig. 3.1 A scenario of cross-technology interference

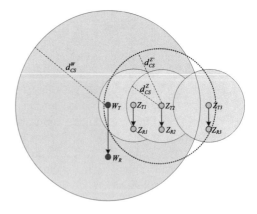

as the two links always have no mutual interference; in the case that a ZigBee transmitter Z_{T2} is within the carrier sense range of WiFi (denoted by d_{CS}^W) but the WiFi transmitter W_T is out of the carrier sense of ZigBee (denoted by d_{CS}^Z), W_T will have severe interference to the ZigBee link $Z_{T2} \longrightarrow Z_{R2}$, as W_T still determines the channel to be idle in this situation due to the low ZigBee signal power, which is below the CCA threshold.

This chapter focuses on the asymmetric CTI scenario and intends to mitigate it through time-domain protocol design. The motivation here is, if the WiFi device can decrease its CCA threshold when it identifies a ZigBee transmission, it can largely avoid CTI through enlarging the ZigBee carrier sense range from d_{CS}^Z to $d_{CS}^{Z'}$, as shown in Fig. 3.1. The ZigBee signal identification should be performed within very short duration, which is smaller than the DIFS (Distributed Interframe Space, 20μs) or a backoff timeslot (9 μs). This requirement can be satisfied by ZShark, the method proposed in Sect. 2 that can make a WiFi device identify the ZigBee signal within several microseconds. Thus, a WiFi device has the opportunity to make proper channel access decisions according to the identified signals .

3.3 System Design

3.3.1 Overview

In this chapter, we propose E-CCA, a time-domain protocol design to combat CTI in the WiFi and ZigBee coexistence networks. E-CCA only has a few changes on the standard CCA. As shown in Fig. 3.2, the standard CCA determines the channel to be busy only when the detected energy is over a CCA threshold β_E. E-CCA adds the signal identification module to the process, as shown in Fig. 3.2. When a WiFi device determines the channel to be idle according to the standard CCA, the signal identification module is further performed to obtain the signal type: the device determines the channel to be busy if a ZigBee signal is identified, otherwise it still determines the channel to be idle.

Since the E-CCA process is performed within DIFS or a backoff timeslot, it is compatible with the standard CSMA/CA process. The WiFi transmission process through E-CCA is as follows: when a device begins to transmit a packet, it first waits

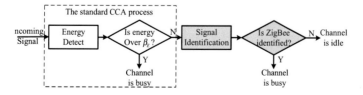

Fig. 3.2 The E-CCA process

for DIFS time; if the channel is determined to be idle through E-CCA during DIFS, the device then waits for a random duration which consists of multiple backoff slots to contend for the channel; the backoff timer is decreased by one when the channel is determined to be idle through E-CCA for a backoff slot, otherwise it is frozen; the device can finally transmit a data packet if the backoff timer reaches zero.

3.3.2 Theoretical Analysis

As shown in Fig. 3.1, E-CCA enlarges the ZigBee carrier sense range d_{CS}^Z to mitigate CTI. Here we investigate the performance improvement of E-CCA theoretically based on the signal identification analysis in Sect. 2.3. We use the results in Table 2.3 since ZShark-CNN has better performance.

According to E-CCA design, the WiFi transmitter W_T shown in Fig. 3.1 needs to identify the ZigBee signal during the ZigBee transmission $Z_{T2} \longrightarrow Z_{R2}$, thus determines the channel to be busy to avoid interference. We see that W_T should determine the channel to be busy when one of the following conditions is satisfied: (1) the sensed signal power at W_T is over the CCA threshold β_E, and (2) the sensed signal power is below β_E but a ZigBee signal can be identified. The two conditions are formulated as follows:

$$(1) P_{BG} + P_r^Z \geq \beta_E,$$
$$(2) P_{BG} + P_r^Z < \beta_E \wedge \frac{P_r^Z}{P_{BG}} > \beta_{SINR_Z} \wedge P_r^Z > P_N, \tag{3.1}$$

where P_r^Z denotes the ZigBee receiving power at the WiFi transmitter, β_{SINR}^Z denotes the SINR threshold over which a ZigBee signal can be identified successfully. Please note that the background interference P_{BG} may be higher than P_N when there are other signals around the two links.

The value of P_r^Z is determined by the 802.11 standard in Condition (1), and is determined by E-CCA in Condition (2). Therefore, the difference of P_r^Z calculated in the two conditions reflects the performance improvement of E-CCA. We let $P_N = -95$ dB and $\beta_E = -62$ dB to make quantitative analysis.

To make quantitative analysis, we let $P_N = -95$ dB and $\beta_E = -62$ dB. The value of β_{SINR_Z} is set to be the minimum $SINR_Z = \frac{P_r^Z}{P_{BG}}$ which makes the signal identification accuracy over 95%, according to Table 2.3. Figure 3.3a shows the changes of P_r^Z when P_{BG} varies from P_N to β_E in the two conditions. We see that E-CCA can decrease the required P_r^Z dramatically nearly in all cases. Specifically, when $P_{BG} = -95$ dB, E-CCA can determine the channel to be busy when P_r^Z is only -93 dB, while that value under standard CCA is up to -63 dB.

It is easy to calculate d_{CS}^Z with P_r^Z in the two conditions. According to the signal propagation model, $P_r = c \cdot \frac{P_t}{d^\alpha}$, where P_t is the transmission power, d is the transmitter-receiver distance, c and α are constant numbers, $\alpha = 3$ when the log

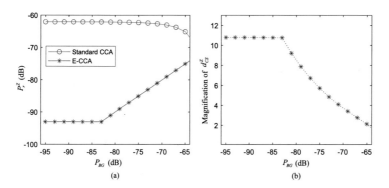

Fig. 3.3 Theoretical analysis of E-CCA performance improvement. (**a**) P_r^z under the two conditions. (**b**) Magnification of $d_c s^z$

distance path loss model is used. Then, $d_{CS}^Z = \sqrt[\alpha]{c \cdot \frac{P_t}{P_r}}$. Here we fix P_t for the ease of analysis, then we have $\frac{d_{CS}^{Z'}}{d_{CS}^Z} = \sqrt[\alpha]{\frac{P_{r1}^Z}{P_{r2}^Z}}$, where P_{r1}^Z and P_{r2}^Z indicate the value P_r^Z calculated by conditions (1) and (2) respectively. We see from Fig. 3.3b that E-CCA can enlarge the ZigBee carrier sense range d_{CS}^Z by more than tenfold when P_{BG} is approximate to P_N. It still has about 4.1× improvement when $P_{BG} = -70\,\text{dB}$.

3.4 Performance Evaluation

In this section, we evaluate the performance improvement of E-CCA in heterogeneous wireless networks through NS-3. We use ZShark-CNN proposed in Sect. 2 here as it has better signal identification performance. We do not implement the details of ZShark-CNN in the simulation, but just utilize the performance investigation results in Table 2.3. During the simulation, when a WiFi device determines the sensed signal is over β_E, it will calculate P_{BG} and $SINR_Z$, then makes the ZigBee signal identified with the corresponding error rate listed in Table 2.3. For example, when P_{BG} is calculated as $-83\,\text{dB}$ while $SINR_Z$ is $-4\,\text{dB}$, the WiFi device will identify a ZigBee signal with the probability of 98.3% if it is in the received signal, and incorrectly identify this signal with the probability of 1.5% if it is not in the received signal.

The basic parameters in the simulations are listed in Table 3.1.

The evaluation is conducted under two scenarios, a two-link topology and a random topology.

Table 3.1 Simulation parameters

parameter	Value	Parameter	Value
WiFi Tx Power	16 dBm	ZigBee Tx Power	0 dBm
WiFi Tx rate	54 Mbps	ZigBee Tx rate	250 Kbps
WiFi slot time	$9\mu s$	ZigBee slot time	$320\mu s$
WiFi SNR threshold	24.5 dB	ZigBee SNR threshold	0 dB

Fig. 3.4 The two-link topology

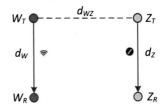

3.4.1 Two-Link Topology

As shown in Fig. 3.4, the two-link topology contains one WiFi link $W_T \rightarrow W_R$ and one ZigBee link $Z_T \rightarrow Z_R$. The link distances of WiFi and ZigBee are denoted by d_w and d_z, respectively. The distance between the two links is denoted by d_{wz}. We investigate the performance of WiFi and ZigBee under different situations through varying the values of d_w, d_z and d_{wz}. We set the WiFi and ZigBee packet delivery rates in the application layer to be 10 Mbps and 100 Kbps, respectively. The other parameters in Table 3.1 are fixed.

The WiFi and ZigBee network throughput in terms of d_w, d_z and d_{wz} are shown in Fig. 3.5. We fix the link distances of d_w and d_z to 5, 10, 20,and 40 m, then adjust d_{wz} for each link distance to investigate the throughput. Simulation results can be summarized into four cases:

Case 1: When the WiFi transmitter W_T is within d_{CS}^Z (about 6 m in this situation), it can determine the channel to be busy and prohibit its own transmissions to avoid interference. E-CCA has the same performance with the standard in this situations, such as $d_{wz} \leq 5$ m in Fig. 3.5,

Case 2: When the WiFi transmitter W_T is out of d_{CS}^Z but within $d_{CS}^{Z'}$, it will interfere with the ZigBee data transmission under the standard, but avoid the interference under E-CCA. E-CCA has about 50% ZigBee throughput improvement in this situation, such as 5 m $< d_{wz} <$ 10 m in Fig. 3.5a, 5 m $< d_{wz} <$ 20 m in Fig. 3.5b and 5 m $< d_{wz} <$ 40 m in Fig. 3.5c, d. We note that $d_{CS}^{Z'}$ is about 35m in this simulation.

Case 3: When the WiFi transmitter W_T is out of both d_{CS}^Z and $d_{CS}^{Z'}$, while its transmission can still interfere with the ZigBee link, E-CCA has similar performance with the standard, such as 40 m $\leq d_{wz} <$ 80 m in Fig. 3.5d.

Case 4: When the WiFi transmission does not interfere with the ZigBee link, we see much higher ZigBee performance under both E-CCA and the standard.

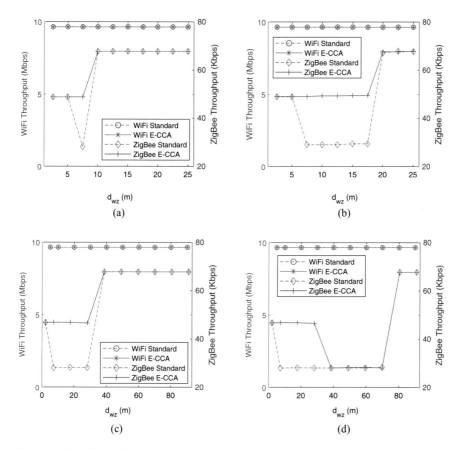

Fig. 3.5 The WiFi and ZigBee throughput under the two-link topology. (**a**) The transmitter-receiver distance is 5 m. (**b**) The transmitter-receiver distance is 10 m. (**c**) The transmitter-receiver distance is 20 m. (**d**) The transmitter-receiver distance is 40 m

Since the typical ZigBee transmission range is as large as ten to hundred meters, E-CCA can improve the network throughput significantly in most cases.

3.4.2   Random Topology

We further evaluate the performance of E-CCA in a random topology. The topology includes two WiFi links and ten ZigBee links. We set the WiFi transmission range to 40 m, and set the ZigBee transmission range to 25 m, according to [6, 11, 14]. All the WiFi and ZigBee links are set up in a 100 × 100 area.

We first change the ZigBee delivery rate and investigate the WiFi and ZigBee throughput. The WiFi delivery rate is fixed to 10 Mbps. From Fig. 3.6a we see

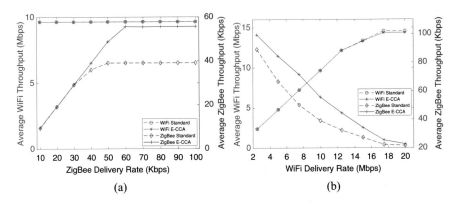

Fig. 3.6 The WiFi and ZigBee throughput under the random topology. (**a**) Throughput in terms of ZigBee delivery rate. (**b**) Throughput in terms of WiFi delivery rate

the increase of ZigBee throughput as the ZigBee delivery rate increases, until reaching a stable value under the saturated situation. In addition, the ZigBee network throughput can be increased by E-CCA through avoiding interference; the throughput gain reaches 43.1% when the ZigBee delivery rate is above about 60 Kbps.

We then change the WiFi delivery rate and investigate the WiFi and ZigBee throughput. The ZigBee delivery rate is fixed to 250 Kbps, while the WiFi physical data rate is 54 Mbps. From Fig. 3.6b we see the increase of WiFi throughput and decrease of ZigBee throughput as the WiFi delivery rate increases. We consider the results come from the shorter WiFi backoff timeslot compared to ZigBee (prease refer to Table 3.1), making the WiFi transmitter have much higher priority to access the channel. Especially, the ZigBee throughput is very poor when the WiFi delivery rate increases to 18 Mbps, as the ZigBee devices have no transmission opportunity in this situation. However, E-CCA can increase the ZigBee throughput significantly under unsaturated WiFi network situations. For example, E-CCA can increase the ZigBee throughput by 34.5% when the WiFi delivery rate is only 10 Mbps.

3.5 Summary

In this chapter, we propose E-CCA to mitigate CTI in heterogeneous wireless networks. E-CCA utilizes ZShark proposed in Chap. 2 to identify the signal type within several microseconds. It includes a time-domain design to enhance the standard CCA process, so as to avoid CTI to ZigBee transmissions. Simulations based on NS-3 show that E-CCA can improve ZigBee network performance dramatically, with little sacrifice to WiFi performance.

References

1. Bozidar, R., Chandra, R., Gunawardena, D.: Weeble: enabling low-power nodes to coexist with high-power nodes in white space networks. In: Proceedings of the ACM CoNEXT (2012)
2. Chae, Y., Wang, S., Kim, S.M.: Exploiting WiFi guard band for safeguarded ZigBee. In: Proceedings of the ACM SenSys (2018)
3. Chen, R., Gao, W.: Enabling cross-technology coexistence for extremely weak wireless devices. In: Proceedings of the IEEE INFOCOM (2019)
4. Chen, W., Yin, Z., He, T.: Global cooperation for heterogeneous networks. In: Proceedings of the IEEE INFOCOM (2020)
5. Chi, Z., Li, Y., Huang, Z., Sun, H., Zhu, T.: Simultaneous bi-directional communications and data forwarding using a single ZigBee data stream. In: Proceedings of the IEEE INFOCOM (2019)
6. Farahani, S.: ZigBee Wireless Networks and Transceivers. Elsevier, Amsterdam (2008)
7. Hermans, F., Rensfelt, O., Voigt, T., Ngai, E., Nordén, L., Gunningberg, P.: SoNIC: classifying interference in 802.15.4 sensor networks. In: Proceedings of the IEEE IPSN (2013)
8. Hithnawi, A., Shafagh, H., Duquennoy, S.: TIIM: technology-independent interference mitigation for low-power wireless networks. In: Proceedings of the IEEE IPSN (2015)
9. Hithnawi, A., Li, S., Shafagh, H., Gross, J., Duquennoy, S.: CrossZig: combating cross-technology interference in low-power wireless networks. In: Proceedings of the ACM/IEEE IPSN (2016)
10. Huang, J., Xing, G., Zhou, G., Zhou, R.: Beyond co-existence: exploiting WiFi white space for Zigbee performance assurance. In: Proceedings of the IEEE ICNP (2010)
11. Internet Access Guide: What is WiFi's maximum range? (2020). https://internet-access-guide.com/what-is-wifi-maximum-range/
12. Li, Y., Chi, Z., Liu, X., Zhu, T.: Chiron: concurrent high throughput communication for IoT devices. In: Proceedings of the ACM MobiSys (2018)
13. Meng, J., He, Y., Zheng, X., Fang, D., Xu, D., Xing, T., Chen, X.: Smoggy-link: fingerprinting interference for predictable wireless concurrency. In: Proceedings of the IEEE ICNP (2016)
14. Newark: Wireless solutions part 4: ZigBee (2022). https://mexico.newark.com/wireless-solutions-part-4-zigbee
15. Wang, Y., Wang, Q., Zeng, Z., Zheng, G., Zheng, R.: Wicop: engineering wifi temporal white-spaces for safe operations of wireless body area networks in medical applications. In: Proceedings of the IEEE RTSS (2011)
16. Yin, Z., Li, Z., Kim, S.M., He, T.: Explicit channel coordination via cross-technology communication. In: Proceedings of the ACM MobiSys (2018)
17. Yu, Z., Li, P., Boano, C.A., He, Y., Jin, M., Guo, X., Zheng, X.: BiCord: bidirectional coordination among coexisting wireless devices. In: Proceedings of the IEEE ICDCS (2021)
18. Zhang, X., Kang, G.S.: Enabling coexistence of heterogeneous wireless systems: case for ZigBee and WiFi. In: Proceedings of the ACM MobiHoc (2011)
19. Zheng, X., Cao, Z., Wang, J., He, Y., Liu, Y.: ZiSense: towards interference resilient duty cycling in wireless sensor networks. In: Proceedings of the ACM SenSys (2014)

Chapter 4
Cross-Technology Interference Management Through Frequency-Domain Design

Abstract In this chapter, we will introduce the cross-technology coexistence problem, and also the motivation of frequency-domain design to combat this problem and improve the network performance. We will then design and implement SledZig to boost cross-technology coexistence for ZigBee devices through both enabling more transmission opportunities and avoiding interference. We will finally implement SledZig on hardware testbed to evaluate its performance.

Keywords Cross-technology interference management · OFDM · QAM

4.1 Introduction

As the two most common wireless technologies in IoT, WiFi and ZigBee inevitably work in the overlapped channels, leading to severe cross-technology coexistence problem [1, 4–6, 10–12, 14–17]. WiFi and ZigBee have asymmetry power levels. The ZigBee signal is always transmitted at less than $1mW$ for energy saving, while the WiFi signal is transmitted at up to $100mW$ for large coverage [2, 3, 7, 13]. Meanwhile, when the devices are contending channel, WiFi has higher priority than ZigBee and can always win the channel for data transmission, due to their MAC layer design [8, 9]. Thus, the WiFi devices induce severe coexistence problems to ZigBee devices, through either prohibiting the ZigBee devices from data transmission or interfering the ongoing ZigBee data transmission.

In this paper, we propose SledZig, a subcarrier-level energy decreasing mechanism on WiFi to boost ZigBee transmission from the frequency-domain design. SledZig is fully compatible with the standard physical and MAC layer processes, and requires no change on commercial WiFi and ZigBee devices. It decreases the WiFi signal energy on the ZigBee channel while keeps the WiFi transmission power unchanged, through exploiting the features of QAM (quadrature amplitude modulation) modulation in WiFi. QAM is a combination of phase and amplitude modulation methods, making the QAM constellation points have different power levels. By inserting extra bits to original WiFi data bits, we let the QAM points in subcarriers overlapped with the ZigBee channel have the lowest power, while

© The Author(s), under exclusive license to Springer Nature Singapore Pte Ltd. 2023   33
J. Yao, K. Wu, *Cross-Technology Coexistence Design for Wireless Networks*,
SpringerBriefs in Computer Science, https://doi.org/10.1007/978-981-99-1670-2_4

those out of the ZigBee channel remain unchanged, leading to up to 14 dB
energy decreasing on the ZigBee channel. With this energy decreasing, the ZigBee
network performance can be improved dramatically through both enabling more
transmission opportunities and avoiding interference.

From the perspective of usage, SledZig is quite simple. With the original data
bits, the WiFi transmitter first inserts extra bits to generate the transmit bits. When
the transmit bits are passed through the standard WiFi transmission process, the
signal energy on the ZigBee channel can be automatically decreased, thus to boost
ZigBee transmissions. Meanwhile, the WiFi receiver can easily obtain the original
data bits through removing the extra bits from the received bits.

We implement SledZig on hardware testbed based on USRP N210 and TelosB
platforms. Experimental results indicate that SledZig can decrease the WiFi signal
power on a ZigBee channel by up to 14 dB. Meanwhile, it can improve the ZigBee
performance dramatically with as low as 6.94% WiFi throughput loss.

4.2 Motivation

Here we first illustrate the background and a cross-technology coexistence problem,
then explain the opportunity on frequency-domain protocol design.

4.2.1 Background

The standard WiFi transmission process is shown in Fig. 4.1. The data bits are first
passed through the channel coding module to combat interference, and transformed
to complex symbols after QAM modulation; the QAM points are then mapped
into OFDM (orthogonal frequency division multiplexing) subcarriers after the S/P
(serial-to-parallel) module, and output as the time-domain OFDM symbols after
IFFT (inverse fast fourier transform) and P/S (parallel-to-serial) processes; each
OFDM symbol is inserted with CP (cyclic prefix) to eliminate the inter-symbol
interference; the signal will finally be transmitted after RF front end.

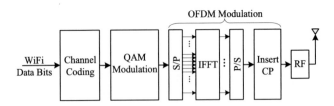

Fig. 4.1 The standard WiFi transmission process

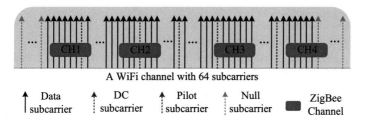

A WiFi channel with 64 subcarriers

↑ Data subcarrier   ↑ DC subcarrier   ↑ Pilot subcarrier   ↑ Null subcarrier   ■ ZigBee Channel

Fig. 4.2 An illustration of the WiFi channel overlapping with four ZigBee channels

Using OFDM, each 20 MHz WiFi channel is divided into 64 subcarriers, including 48 data subcarriers, 4 pilot subcarriers and 12 null subcarriers, as shown in Fig. 4.2. One WiFi channel overlaps with four ZigBee channels, we call them as CH1, CH2, CH3 and CH4 for the ease of description. Actually, each WiFi channel overlaps with four ZigBee channels in the same way as that in Fig. 4.2. Among the four channels, CH1, CH2 and CH3 overlap with a pilot subcarrier while CH4 overlaps with null subcarriers.

4.2.2 Cross-Technology Coexistence Problem

As analysis in Sect. 1.2.1 that, WiFi and ZigBee are different in both physical and MAC specifications. Actually, the asymmetry transmission power and MAC parameters lead to severe CTI to ZigBee transmissions, falling into two scenarios.

The first scenario is that the WiFi transmission always prohibits some ZigBee transmissions due to the larger WiFi carrier sense range d_{CS}^W. We see from Fig. 4.3a

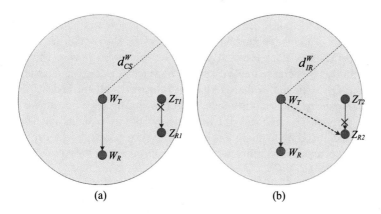

(a)   (b)

Fig. 4.3 Two scenarios that WiFi affects the ZigBee performance. (**a**) ZigBee devices within the WiFi carrier sense range d_{cs}^w are prohibited to transmit data. (**b**) The ZigBee transmission interfered by the WiFi transmission

that, under the coexistence of $W_T \longrightarrow W_R$ and $Z_{T1} \longrightarrow Z_{R1}$, the ZigBee link is always prohibited from data transmission, due to the unfairness in channel competition. According to WiFi and ZigBee CSMA/CA parameters shown in Table 3.1, the DIFS or backoff timeslot duration of WiFi is much shorter than that of ZigBee. Therefore, when both the WiFi transmitter W_T and the ZigBee transmitter Z_{T1} contend the channel for data transmission, W_T can always win. Our performance evaluation in Chap. 3 reveals that the ZigBee link can only proceed its data transmission only the WiFi link is very unsaturated (the WiFi application layer data rate should be below 20% of the physical layer data rate).

The second scenario is that the WiFi transmission always interferes with the ZigBee transmission due to the larger WiFi interference range d_{IR}^W. We see from Fig. 4.3b that, during the data transmission of $Z_{T2} \longrightarrow Z_{R2}$, this ZigBee link has a high probability to be interfered by the strong WiFi signal from W_T when it is within d_{IR}^W.

4.2.3 Opportunity of Frequency-Domain Design

Based on the aforementioned analysis, we see that the ZigBee network performance can be obviously increased when the WiFi transmission power is decreased. In the scenario of Fig. 4.3a, the ZigBee transmitter Z_{T1} has the opportunity to be out of the shortened d_{CS}^W and can transmit data to Z_{R1}; in the scenario of Fig. 4.3b, the ZigBee transmission $Z_{T2} \longrightarrow Z_{R2}$ has the opportunity to proceed successfully due to the decreased WiFi interference.

However, how to decrease the WiFi signal power on the ZigBee channel without affecting the WiFi performance is a big challenge. Our observation on the WiFi OFDM modulation reveals that the WiFi signal power on the overlapped subcarriers can be decreased through designing low power constellation points. According to the WiFi transmission process shown in Fig. 4.1, the QAM modulation is conducted before the OFDM module. Each QAM constellation point carry information through both amplitude and phase modulations. Figure 4.4a shows the 16 constellation points of QAM-16, we see four red points have the lowest power. Thus, we have the opportunity do decrease the WiFi signal power on the ZigBee channel through making the QAM points in the overlapped subcarriers all the red ones, and the expected frequency spectrum is shown in Fig. 4.4b. This idea has little impact on the WiFi performance as the WiFi transmission power remains unchanged. We adopt payload encoding to attain this goal in the system design in order to achieve commercial device compatibility, which will be discussed in the following parts.

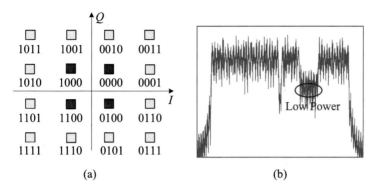

Fig. 4.4 An example of the QAM-16 lowest points and the frequency spectrum when all the overlapped subcarriers are filled with the lowest points. (**a**) QAM-16 constellation points. (**b**) WiFi frequency spectrum

4.3 System Design

In this chapter, we propose SledZig design to improve the ZigBee network performance. SledZig includes the design at WiFi transmitter and receiver side. No changes are required at ZigBee devices.

4.3.1 System Design at WiFi Transmitter

The system design at the WiFi transmitter side is shown in Fig. 4.5. The payload encoding module inserts extra bits to the WiFi data bits to generate the transmit bits. When the transmit bits are passed through the WiFi transmission process including

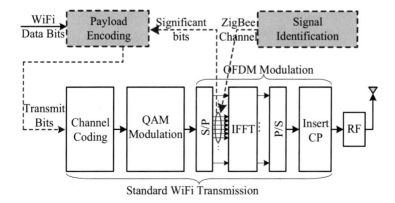

Fig. 4.5 The SledZig design at WiFi transmitter side

channel coding, QAM modulation and OFDM modulation, the QAM points in the overlapped subcarriers are with the lowest power, so as to decrease the WiFi signal power on the ZigBee channel. Please note that the ZigBee channel should be obtained through the signal identification process in Chap. 1; as each WiFi channel overlaps with four ZigBee channel in the same pattern shown in Fig. 4.2, it is easy to obtain the overlapped OFDM subcarriers with the identified ZigBee channel.

In the following parts, we will give the detailed design of payload encoding following the WiFi transmission process.

4.3.1.1 Significant Bits

Based on the SledZig design, the QAM points filled in the overlapped subcarriers should have the lowest power. However, not all the bits in each point are significant. As shown in Fig. 4.3a, one QAM-16 point carries four bits, and only two of them are significant to make the power lowest. We call them significant bits, as the shadowed ones shown in Table 4.1. Similarly, there are four and six significant bits in each QAM-64 and QAM-256 points. The payload encoding module only needs to insert extra bits to make the significant bits be the designated ones.

4.3.1.2 Channel Coding

The key issue in SledZig design is how to insert extra bits to the original WiFi data bits to generate significant bits in corresponding positions. Channel coding brings a big challenge for this process. This part focuses on payload encoding design through following the reverse channel coding process step by step.

As shown in Fig. 4.6, channel coding in WiFi transmission includes interleaver, convolutional encoder and scrambler. Interleaver is used in wireless communication system to reduce the decoding errors, while scrambler is used to avoid long sequences of bits with the same value. Both modules are one-by-one mapping from input bits to output bits. The convolutional encoder adds redundancy to the data bits, and it cannot generate arbitrary bit sequence.

With the identified ZigBee channel and QAM modulation type, it is easy to get the significant bits before QAM modulation, and get those before interleaving through one-by-one mapping. We denote the significant bits before interleaver as

Table 4.1 The significant bits

Bits	QAM-16				QAM-64						QAM-256							
	0	0	0	0	0	1	0	0	1	0	0	1	0	0	0	1	0	0
	0	1	0	0	0	1	0	1	1	0	0	1	0	0	1	1	0	0
	1	0	0	0	1	1	0	0	1	0	1	1	0	0	0	1	0	0
	1	1	0	0	1	1	0	1	1	0	1	1	0	0	1	1	0	0

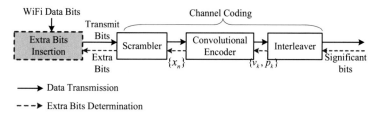

Fig. 4.6 The payload encoding process

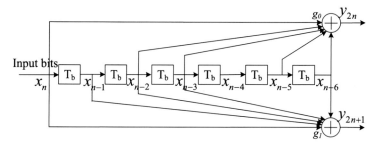

Fig. 4.7 The process of 1/2-rate convolutional encoding

$\{v_k, p_k\}(k \in [1, K])$, where v_k and p_k indicate the value and position of the k-th significant bit. Then, the key process here is to determine the scrambled transmit bits $\{x_n\}\}$ by $\{v_k, p_k\}(k \in [1, K])$. We achieve this goal through analyzing the convolutional encoding process, summarizing its characteristic to determine the extra bits.

The 802.11 standard recommends several coding rates under each QAM modulation to achieve different WiFi data rates. The basic process is 1/2-rate encoding, while other coding rates are achieved by omitting some of the 1/2 coded bits. Figure 4.7 shows the 1/2-rate encoding process. It adopts two generator polynomials $g_0 = (1011011)_2$ and $g_1 = (1111001)_2$ to generate two coded bits y_{2n-1} and y_{2n} for each input bit x_n. We see from the figure that y_{2n-1} and y_{2n} are determined by a sequence of bits from x_n to x_{n-6}. We let $X_n = [x_n \ x_{n-1} \ x_{n-2} \ x_{n-3} \ x_{n-4} \ x_{n-5} \ x_{n-6}]'$ for the easy of description. Then the one step encoding process to generate two output bits y_{2n-1} and y_{2n} for x_n can be formulated as:

$$g_0 \times_{GF(2)} X_n = y_{2n-1},$$
$$g_1 \times_{GF(2)} X_n = y_{2n}, \tag{4.1}$$

where $GF(2)$ means the calculation is in the Galois Field GF(2).

With significant bits $\{v_k, p_k\}$ after encoder, then the extra bits in the uncoded bits $\{x_n\}$ can be determined through solving Eq. (4.1) one by one. Table 4.2 lists the 14 significant bits in the first OFDM symbol under QAM-16 and the ZigBee channel of CH2. Please note that the values of v_k are always zero in this situation, so they are

Table 4.2 An example of significant bits in the first OFDM symbol

k	1	2	3	4	5	6	7	8	9	10	11	12	13	14
p_k	29	30	41	42	77	78	89	90	125	138	172	173	183	186
n	15	15	21	21	39	39	45	45	63	69	86	87	92	93

not shown in this table. We also give the corresponding index n for each significant bit, which means, the significant bit in the position p_k is triggered by the n-th input bit.

The extra bits insertion process based on significant bits can be divided into two situations. The first situation is that, given a n, only one of the two bits y_{2n-1} and y_{2n} in Eq. (4.1) is a significant bit, such as the case of $k = 9$, where $n = 63$ and $p_k = 2n - 1 = 125$ in Table 4.2. We call this kind of bits as *single significant bit*. The second situation is that, both the two bits y_{2n-1} and y_{2n} are significant bits, such as the case of $k = 1$ and $k = 2$, where $n = 15$. We call this kind of bits as *twin significant bits*.

The extra bits insertion in the *single significant bit* situation is very simple. We let x_n be the extra bit and all the previous bits from x_{n-1} to x_{n-6} are known bits; then x_n can be obtained easily through solving the corresponding equation in 4.1. The extra bits insertion in the *twin significant bit* situation is very much more complex, as Eq. (4.1) can only be satisfied with two unknowns, corresponding to two extra bits. We let x_n and x_{n-1} be the extra bits, while x_{n-1} is also used to generate the previous coded bits from $y_{2(n-1)-1}$ to $y_{2(n-1)}$. Once there are additional significant bits among them, Eq. (4.1) may not be satisfied. However, we find this case does not occur in the whole extra bits determination process, as the deinterleaving process has scattered the significant bits far way enough to avoid this case, no matter in which combination of QAM modulations and ZigBee channels.

The general extra bits insertion process is formulated in Algorithm 1. $\{x_i'\}$ and $\{x_n\}$ represent the scrambled WiFi data bits and the scrambled transmit bits, respectively. The final transmit bits are finally calculated through descrambling $\{x_n\}$. From the first bit in $\{x_i'\}$, the device determines whether it triggers a significant bit. If yes, it calculates the extra bits etr_0 or etr_1, then adjusts the values of $\{x_n\}$; if not, it simply assigns current x_i' to x_n. The process is conducted until all the data bits $\{x_i'\}$ are traversed.

4.3.1.3 Impact of Pilot

There are four pilot sucarriers in each WiFi channel, as shown in Fig. 4.2. The pilot can not be changed according to the commercial WiFi transmission process, while three ZigBee channels (CH1, CH2 and CH3) inevitably overlap with a pilot subcarrier. With much higher power, the pilot will obviously decrease the SledZig performance due to the increased signal power at ZigBee. However, as long as the averaged WiFi singal power is decreased to make the ZigBee SNR (signal

---

Algorithm 1: Transmit bits generation process

Input   : Data bits $\{x_i'\}, i \in [1, N']$;
                 Significant bits $\{v_k, p_k\}, k \in [1, K]$.
Output: Transmit Bits $\{x_n\}, n \in [1, N]$.

1 $k \leftarrow 1$; $n \leftarrow 1$; $etr_0 \leftarrow 0$; $etr_1 \leftarrow 0$; $tmp \leftarrow 0$.
2 **while** $i \leq N'$ **do**
3   **if** $(2n + 1)==p_k$ *and* $2n + 2==p_{k+1}$ **then**
4            $X_{n+1} = [etr_0 \; etr_1 \; x_{n-1} \; x_{n-2} \; x_{n-3} \; x_{n-4} \; x_{n-5}]'$;
5            $y_{2n+1} \leftarrow v_k, y_{2n+2} \leftarrow v_{k+1}$;
6   Calculate etr_0 and etr_1 through Eq. (4.1);
7            $x_n \leftarrow etr_1$;
8            $n \leftarrow n + 1$;
9            $x_n \leftarrow etr_0$;
10           $n \leftarrow n + 1; k \leftarrow k + 2$.
11   **else if** $(2n - 1)==p_k$ *or* $2n==p_k$ **then**
12           $X_n = [etr_0 \; x_{n-1} \; x_{n-2} \; x_{n-3} \; x_{n-4} \; x_{n-5} \; x_{n-6}]'$;
13   **if** $(2n - 1)==p_k$ **then**
14                $y_{2n-1} \leftarrow v_k$.
15   **else**
16                $y_{2n} \leftarrow v_k$.
17   Calculate etr_0 through Eq. (4.1);
18           $x_n \leftarrow etr_0$;
19           $n \leftarrow n + 1; k \leftarrow k + 1$.
20   **else**
21           $x_n \leftarrow x_i'$;
22           $i \leftarrow i + 1$;
23           $n \leftarrow n + 1$.

---

to noise ratio) meet the decoding requirement, the ZigBee transmission can still be successful. This result benefits from DSSS (Direct Sequence Spread Spectrum) modulation adopted by ZigBee. DSSS can resist high interference within very short frequency band, while here the pilot subcarrier occupies only 312.5 KHz band, which is much narrower than the 2 MHz ZigBee channel.

4.3.1.4   Impact of WiFi Preamble

The WiFi preamble also cannot be changed according to the previous design. It contains 10 repetitive STS (short training symbols) as well as two repetitive LTS (long training symbols), and lasts for 16 μs, as shown in Fig. 4.8. Meanwhile, each ZigBee packet includes a preamble with the duration of 128 μs before the payload, and the payload contains a set of ZigBee symbols to carry data bits. Each ZigBee symbol lasts for 16 μs. Here the impact of WiFi preamble is analyzed from the two scenarios shown in Fig. 4.3.

Fig. 4.8 The packet structure

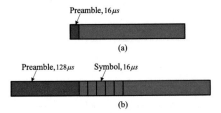

For the scenario of Fig. 4.3a where SledZig shortens d_{CS}^{W} to enable more ZigBee transmissions, the impact is very small, as the ZigBee CCA lasts for 128 μs, which is much longer than the WiFi preamble. In this case, the 16 μs high power signal will be averaged by the remaining 112 μs low power signal, thus have little impact on the CCA result.

For the scenario of Fig. 4.3b where SledZig reduces the WiFi interference to ZigBee transmission, the impact has two situation. If the ZigBee preamble is interfered, its redundancy design can resist this kind of interference, making the impact negligible. If the ZigBee payload is interfered, there is a high probability that the corresponding ZigBee symbols cannot to be detected correctly. However, we will show in Sect. 4.4 that SledZig can still improve the ZigBee network performance significantly with this limitation.

4.3.2   System Design at WiFi Receiver

The process at the WiFi receiver side is quite simple: the receiver first conducts the standard WiFi receiving process to obtain the transmit bits, then removes the extra bits to get the original WiFi data bits. The positions of extra bits are fixed in the transmit bits, and they are determined by three kinds of information: the ZigBee channel, QAM modulation and coding rate. The latter two information can be obtained directly from the PLCP (physical layer convergence protocol) header of the WiFi packet [8]. The key issue here is to obtain the ZigBee channel. With the transmit bits, the WiFi receiver can conduct the channel coding and modulation process shown in Fig. 4.3, then it can observe the QAM points and determine the ZigBee channel: the QAM points in the overlapped subcarriers are all lowest ones. This process is fully compatible with the 802.11 standard.

4.4   Performance Evaluation

In this section, we evaluate the ZigBee and WiFi network performance through hardware experiments.

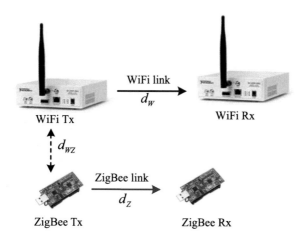

Fig. 4.9 Experimental setup

4.4.1 Experimental Setup

We implement a testbed based on USRP (universal software radio peripheral) N210 and TelosB for performance evaluation, as shown in Fig. 4.9. We let one USRP (WiFi Tx) mimic the WiFi transmitter and generate the WiFi signals following IEEE 802.11 standard, while the transmit bits are generated offline through encoding the original payload. We let another USRP (WiFi Rx) mimic the WiFi receiver to demodulate the transmit bits. Two TelosB platforms are used as the ZigBee transmitter and receiver to test the ZigBee performance.

We conduct the experiments in a $10 \times 15\,m$ open space office, with the tested background noise of $-91\,dB$. The USRP Tx and Rx operate on the 13th WiFi channel, while the TelosB platforms operate on one of the four overlapped ZigBee channels numbering from 23 to 26, corresponding to CH1 to CH4 in Fig. 4.2.

As shown in Fig. 4.9, to simplify the description in following parts, we use d_W, d_Z, and d_{WZ} to represent the link distance between WiFi Tx and Rx, the link distance between ZigBee Tx and Rx, and the distance between WiFi and ZigBee links, respectively.

4.4.2 Performance of ZigBee Transmission

4.4.2.1 RSSI at ZigBee

Since SledZig intends to decrease the WiFi signal power on the ZigBee channel to improve the ZigBee performance, we first investigate how much signal power can be reduce, using RSSI (received signal strength indication) collected by TelosB.

Figure 4.10 shows the decrease of RSSI under different QAM modulations and ZigBee channels. The normal WiFi signal leads to little change on RSSI under different QAM modulation types due to similar averaged signal power. Meanwhile, the RSSI values collected on CH1, CH2 and CH3 are comparable because all of them are overlapped with one pilot subcarrier as well as seven data subcarriers. However, we see about 3~4 dB lower RSSI collected on CH4, due to the two overlapped null subcarriers. From CH1 to CH3, SledZig can decrease RSSI from about −60 dB with normal WiFi signal, to −64, −66, and −68 dB under QAM-16, QAM-64 and QAM256, respectively. In CH4, RSSI can be further decreased from about −64 to −70, −75, and −78 dB, respectively. The results show that pilot overlapped with CH1~CH3 has a great impact to the averaged signal power, and a ZigBee network can have the highest performance when it works on CH4.

4.4.2.2 Impact of d_{WZ}

We then investigate the ZigBee performance in terms of d_{WZ} under continuous WiFi transmissions. We set the WiFi Tx gain to 15 dB, set the ZigBee link distance d_Z to $1m$. The ZigBee throughput under three QAM modulations is shown in Fig. 4.11. SledZig can make ZigBee transmission successful under much shorter d_{WZ} values. When ZigBee operates on the channels from CH1 to CH3, the d_{WZ} value which make ZigBee transmission successful under normal WiFi signal is about 8.5 m; however, under SledZig with QAM-16, QAM-64 and QAM-256, the values are shortened to 5 m, 4.5 m and 3.5 m, respectively. Please note that the ZigBee throughput without interference is only about 63 Kbps, which is much lower than the 250 Kbps data rate in the physical layer. We consider that is mainly because of the long duration of backoffs and DIFS in the MAC layer design. When ZigBee operates on the channel of CH4 as shown in Fig. 4.11b, SledZig can make Zigbee transmission successful under QAM-256 even when d_{WZ} is as short as 1 m, as the WiFi signal power in this channel is 4 dB lower than that in CH1-CH3. The results indicate that SledZig makes the ZigBee links have more chances to proceed successfully, while the benefit comes from the shortened WiFi carrier sense range for ZigBee (d_{CS}^W in Fig. 4.4a).

4.4.2.3 Impact of d_Z

We then investigate the ZigBee performance in terms of d_Z under continuous WiFi transmissions. In this experiments, we set the ZigBee channel to CH4, and set d_{WZ} to 6 m so that the ZigBee Tx is out of d_{CS}^W and can transmit packets with both normal and SledZig signals. From the experimental results in Fig. 4.12 we see that,

Fig. 4.10 RSSI collected at ZigBee under each channel and modulation type. (**a**) QAM-16. (**b**) QAM-64. (**c**) QAM-256

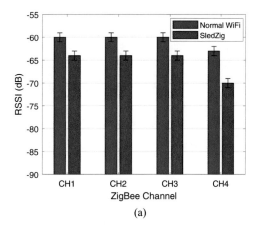

(a)

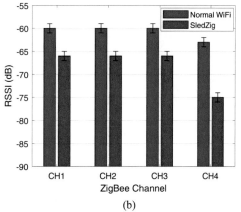

(b)

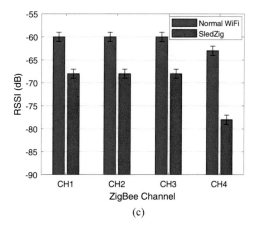

(c)

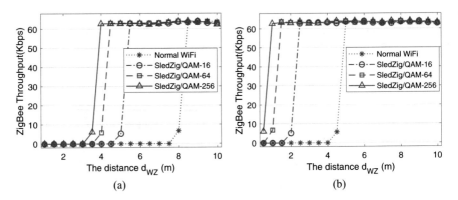

Fig. 4.11 The ZigBee throughput in terms of d_{WZ} under continuous WiFi transmission. (**a**) CH1–CH3. (**b**) CH4

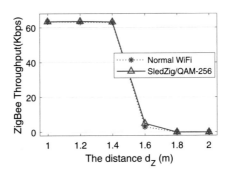

Fig. 4.12 The ZigBee throughput in terms of d_Z under continuous WiFi transmission

the ZigBee throughput is nearly zero when d_Z increases to 1.6 m, due to the weak ZigBee signal at the receiver side, no matter using normal WiFi or SledZig. SledZig has poor performance in this case mainly due to the WiFi preamble, which inevitably overlaps with the ZigBee payload as the WiFi packets are transmitted continuously. The high power of the WiFi preamble induces severe interference to the ZigBee packets, leading to low throughput. However, we will conduct experiments in the following part to illustrate that SledZig can still improve the ZigBee throughput with lower WiFi traffic.

4.4.2.4 Impact of WiFi Traffic

We further conduct experiments under lower WiFi traffic to investigate the ZigBee throughput. We set the ZigBee channel to CH3, set d_{WZ} and d_Z to 1 and 0.5 m, respectively. The ZigBee link in this situation has a high probability to be interfered by the WiFi signal under continuous WiFi transmissions. Here we change the duration ratio, the ratio of the WiFi transmission duration to the test duration, to evaluate the ZigBee throughput. Duration ratio reflects the amount of data in the application layer. Figure 4.13 shows the results using box plots as the throughput fluctuates in a large range due to random interference situations. We see significant ZigBee throughput improvement under low data traffic. Specifically, the throughput is about 23 Kbps under normal WiFi signal only when the ratio is equal to or

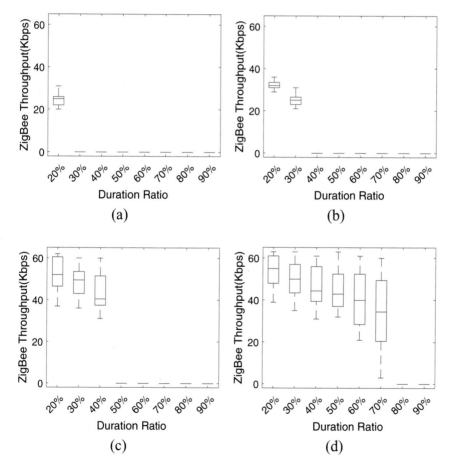

Fig. 4.13 The ZigBee throughput under different WiFi data traffic. (**a**) Normal WiFi. (**b**) SledZig/QAM-16. (**c**) SledZig/QAM-64. (**d**) SledZig/QAM-256

less than 20%, while it is increased to about 50 Kbps even when the ratio is 70% under SledZig with QAM-256. The main reason is that, under the low WiFi traffic situations, the preamble in SledZig signals may partially interfere with the ZigBee payload, and partially interfere with the ZigBee preamble, while the latter does not affect the ZigBee packet reception.

4.4.3   Performance of WiFi Transmission

The design of SledZig will obviously degrade the WiFi network performance since it inserts extra bits to the WiFi data bits. Here we mainly investigate the WiFi throughput loss.

Table 4.3 shows the modulation types and coding rates recommended by the IEEE 802.11 standard, as well as the number of extra bits in one OFDM symbol under each setting. The number of extra bits is only determined by the positions of significant bits, which are affected by QAM modulation and ZigBee channel. The coding rate does not affect this number as all the other coding rates are achieved based on the 1/2-rate encoding.

Table 4.3 also shows the WiFi throughput loss of SledZig under each setting, ranging from 6.94% to 14.58%. We see the highest loss of 14.58% under the settings of ZigBee channels from CH1 to CH3, and QAM-16 with 1/2-rate encoding, QAM-64 with 2/3-rate encoding, or QAM-256 with 3/4-rate encoding. The lowest throughput loss of 6.94% is achieved under ZigBee channel of CH4 and QAM-16 with 2/3-rate encoding. The throughput loss under CH4 is lower than that under CH1-CH3, due to fewer extra bits.

4.5   Summary

In this chapter, we propose SledZig to enable coexistence of heterogeneous wireless devices from the frequency-domain design, so as to improve the ZigBee network performance. SledZig decreases the WiFi signal power on the ZigBee channel through payload encoding, which insert extra bits to the WiFi data bits to make QAM points in the overlapped subcarriers have the lowest power. Experimental results based on hardware testbed show that SledZig can improve ZigBee throughput significantly with little sacrifice on WiFi throughput.

Table 4.3 The SledZig parameters and WiFi throughput loss under different settings

Modulation	Coding rate	No. of bits per OFDM symbol	No. of extra bits (CH1-CH3)	No. of extra bits (CH4)	Throughput loss (CH1-CH3)	Throughput loss (CH4)
QAM-16	1/2	96	14	10	14.58%	10.42%
	2/3	144	14	10	9.72%	6.94%
QAM-64	2/3	192	24	20	14.58%	10.42%
	3/4	216	28	20	12.96%	9.26%
	5/6	240	28	20	11.67%	8.33%
QAM-256	3/4	288	42	30	14.58%	11.72%
	5/6	320	42	30	13.12%	9.37%

References

1. Božidar, R., Chandra, R., Gunawardena, D.: Weeble: enabling low-power nodes to coexist with high-power nodes in white space networks. In: Proceedings of the ACM CoNEXT (2012)
2. Chae, Y., Wang, S., Kim, S.M.: Exploiting WiFi guard band for safeguarded ZigBee. In: Proceedings of the ACM SenSys (2018)
3. Chen, R., Gao, W.: Enabling cross-technology coexistence for extremely weak wireless devices. In: Proceedings of the IEEE INFOCOM (2019)
4. Chen, W., Yin, Z., He, T.: Global cooperation for heterogeneous networks. In: Proceedings of the IEEE INFOCOM (2020)
5. Hermans, F., Rensfelt, O., Voigt, T., Ngai, E., Nordén, L., Gunningberg, P.: SoNIC: classifying interference in 802.15.4 sensor networks. In: Proceedings of the IEEE IPSN (2013)
6. Hithnawi, A., Shafagh, H., Duquennoy, S.: TIIM: technology-independent interference mitigation for low-power wireless networks. In: Proceedings of the IEEE IPSN (2015)
7. Hithnawi, A., Li, S., Shafagh, H., Gross, J., Duquennoy, S.: CrossZig: combating cross-technology interference in low-power wireless networks. In: Proceedings of the ACM/IEEE IPSN (2016)
8. IEEE Computer Society. 802.11: Wireless LAN medium access control (MAC) and physical layer (PHY) specifications amendment 5: enhancements for higher throughput (2009)
9. IEEE Computer Society. 802.15.4: IEEE standard for low-rate wireless networks (2016)
10. Li, Y., Chi, Z., Liu, X., Zhu, T.: Chiron: concurrent high throughput communication for IoT devices. In: Proceedings of the ACM MobiSys (2018)
11. Meng, J., He, Y., Zheng, X., Fang, D., Xu, D., Xing, T., Chen, X.: Smoggy-link: fingerprinting interference for predictable wireless concurrency. In: Proceedings of the IEEE ICNP (2016)
12. Wang, Y., Wang, Q., Zeng, Z., Zheng, G., Zheng, R.: Wicop: engineering wifi temporal white-spaces for safe operations of wireless body area networks in medical applications. In: Proceedings of the IEEE RTSS (2011)
13. Yao, J., Huang, H., Xie, R., Zheng, X., Wu, K.: SledZig: boosting cross-Technology coexistence for low-Power wireless devices. In: Proceedings of the IEEE ICDCS (2022)
14. Yao, J., Lou, W., Xie, R., Jiao, X., Wu, K.: Mitigating cross-technology interference through fast signal identification. IEEE Trans. Vehi. Tech. **72**(2), 2521–2534 (2023)
15. Yin, Z., Li, Z., Kim, S.M., He, T.: Explicit channel coordination via cross-technology communication. In: Proceedings of the ACM MobiSys (2018)
16. Yu, Z., Li, P., Boano, C.A., He, Y., Jin, M., Guo, X., Zheng, X.: BiCord: bidirectional coordination among coexisting wireless devices. In: Proceedings of the IEEE ICDCS (2021)
17. Zhang, X., Kang, G.S.: Enabling coexistence of heterogeneous wireless systems: case for ZigBee and WiFi. In: Proceedings of the ACM MobiHoc (2011)

Chapter 5
Cross-Technology Communication Through Symbol-Level Energy Modulation

Abstract The coexistence of heterogeneous devices in wireless networks brings a new topic on cross-technology communication (CTC) to improve the coexistence efficiency and boost collaboration among these devices. In this chapter, we will propose symbol-level energy modulation (SLEM) to achieve WiFi to ZigBee CTC transmission in commercial wireless devices. We will also implement SLEM on hardware testbed to evaluate its performance.

Keywords Cross-technology communication · OFDM · QAM

5.1 Introduction

Cross-technology communication (CTC), which establishes direct communication among heterogeneous devices [2, 8–10, 15], has the potential to bring about quite a few benefits and applications [8, 12, 13], such as combating the cross-technology interference through exchanging coordination information among the devices [12], enabling the WiFi AP to directly control the Zigbee devices deployed for smart home [8], and etc. "

Current works on WiFi to ZigBee CTC design are generally achieved through two methods: *physical-layer CTC* and *packet-level energy modulation* (PLEM). The *physical-layer CTC* makes a commercial WiFi device transmit ZigBee signals directly through signal emulation, such that this signal can be detected through ZigBee normal demodulation process [9]. It achieves the high CTC data rate comparable to a ZigBee radio. The main problem is that it is hard to be deployed in commercial wireless networks due to the channel incompatible. The PLEM methods convey cross-technology information through employing the packet-level features, like packet transmission duration [1, 4], duration pattern [7, 11], and interval [8, 14], so that receivers can detect the information through energy sensing. This kind of methods are compatible with commercial devices in channel and the physical layer process. However, they have the drawbacks of low CTC data rate and MAC incompatibility with commercial devices.

J. Yao, K. Wu, *Cross-Technology Coexistence Design for Wireless Networks*,
SpringerBriefs in Computer Science, https://doi.org/10.1007/978-981-99-1670-2_5

In this chapter, we propose symbol-level energy modulation (SLEM) to achieve WiFi to ZigBee CTC transmission through payload encoding. In the WiFi transmission process, the extra bits inserted to the WiFi data bits make the QAM points in overlapped subcarriers have highest or lowest power, thus to deliver the CTC information. Experiments conducted on hardware testbed indicate that SLEM can achieve a robust and fast concurrent transmissions of both CTC and WiFi.

5.2 Motivation

This section illustrates the motivation of SLEM through observing on both WiFi and ZigBee transmission processes.

The WiFi packet transmission duration can be up to about 5.48 ms according to the IEEE 802.11 standard, as shown in Table 5.1. Although the physical layer data rate has been increased from 11 Mbps in 802.11b to more than 6 Gbps in 802.11ac [3], the maximum packet transmission duration remains stable due to the MAC layer design, such as using A-MPDU (Aggregated—MAC Protocol Data Unit) to accomplish the super-length packet. Comparing with the 5.48*ms* of the WiFi packet duration, ZigBee devices have much shorter RSSI sampling interval, which is 32 μs for TelosB [15]. Thus, a WiFi transmitter has the opportunity to convey CTC information through making the transmitted packet have a sequence of distinguishable energy, while a ZigBee device can obtain the energy changes through RSSI sampling.

We then investigate how to achieve energy modulation within one WiFi packet. As shown in Fig. 4.1, the WiFi data bits are transformed to QAM points, then mapped into OFDM (orthogonal frequency division multiplexing) subcarriers for transmission. Each QAM point has four points with the lowest power, and four points with the highest power, as shown in Fig. 5.1a which exhibits points for QAM-16. We see that the four blue points have 9× higher energy than the four red points. Thus, we have the opportunity of symbol-level energy modulation using a single WiFi packet. As shown in Fig. 5.1b, if we let the red points carry the CTC information '0' and let the blue points carry '1', the two kinds of information will possess distinguishable energy levels and then have the possibility to be discerned at ZigBee.

Table 5.1 Attribute comparisons of the IEEE 802.11 family

Attribute	802.11a/g	802.11n	802.11ac
Maximum R_w	54 Mbps	600 Mbps	6.9 Gbps
Maximum L_w	4095 bytes	65,535bytes	4,692,480 bytes
Maximum τ_w	5.46 ms	5.484 ms	5.484 ms

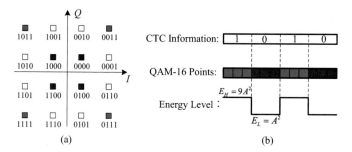

Fig. 5.1 An example of symbol-level energy modulation. (**a**) QAM-16 constellation points. (**b**) Energy modulation for CTC

5.3   System Design

This chapter proposes SLEM to achieve WiFi to ZigBee CTC transmission. In this section, we first give the architecture of SLEM, then give the detailed design at the WiFi transmitter side, the ZigBee receiver side and the WiFi receiver side, respectively.

5.3.1   System Design at WiFi Transmitter

5.3.1.1   Overview

The system design is shown in Fig. 5.2. The WiFi device first performs payload encoding to generate SLEM bits according to the WiFi data bits and significant bits, which are determined by CTC bits and the ZigBee channel. Here the ZigBee channel should be obtained at first through the signal identification process in Chap. 2, so as to determine the OFDM subcarriers to carry the CTC information.

When the SLEM bits are passed through the standard WiFi transmission process, the constellation points within the overlapped subcarriers will carry the CTC information through energy modulation, as shown in Fig. 5.2b; that means, the CTC bit '1' will be carried by the blue QAM points with highest power, and the CTC bit '0' will be carried by the red QAM points with lowest power. Through this way, the transmitted SLEM signal contains both the desired energy modulated CTC signal and the WiFi signal, thus can deliver both kinds of data bits concurrently.

There are two key issues here. The first issue is to determine the CTC symbol duration τ_{CTC}, which contains multiple OFDM symbols with the duration of $\tau_{OFDM} = 4\,\mu s$, as shown in Fig. 5.2b. This parameter will finally determine the significant bits to generate SLEM bits. The second issue is payload encoding process. We will illustrate them in the following parts.

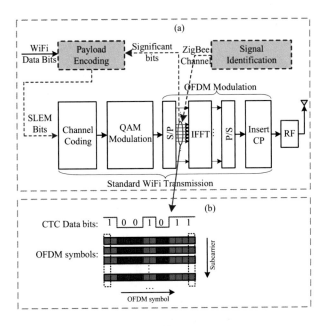

Fig. 5.2 SLEM design at the WiFi transmitter side

5.3.1.2   CTC Symbol Duration Determination

The CTC symbol duration τ_{CTC} is decided by the characteristic of RSSI sampling at the ZigBee receiver. We use TelosB as the ZigBee platform in this book. The RSSI samples are generated every 32 μs, while the values are averaged over 128 μs. We find that both τ_{CTC} and the RSSI sample positions will affect the RSSI values, as shown in Fig. 5.3. The CTC bits $\{1, 0, 1, 1\}$ are carried by a sequence of QAM points with energy $\{E_H, E_L, E_H, E_H\}$, where E_H and E_L indicate the high and

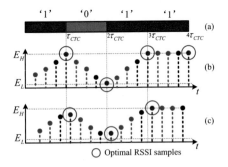

Fig. 5.3 An example of RSSI sampling at the receiver side. (**a**) Transmit CTC symbols. (**b**) RSSI values when sampled at CTC symbol boundaries. (**c**) RSSI values when **not** sampled at CTC symbol boundaries

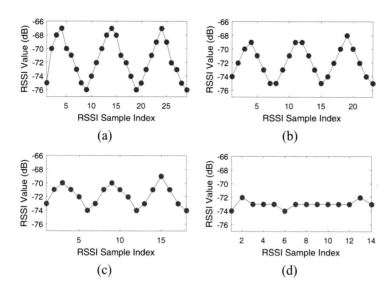

Fig. 5.4 The RSSI samples under different CTC symbol durations through experiments. (**a**) τ_{CTC}=160 μs. (**b**) τ_{CTC}=128 μs. (**c**) τ_{CTC}=96 μs. (**d**) τ_{CTC}=64 μs

low energy levels, and $E_H = 9 \times E_L$ under QAM-16. Here $\tau_{CTC} = 128$ μs, and the RSSI values are obtained every 32 μs. We see very different RSSI values even within one CTC symbol, and only the RSSI values at the CTC symbol boundaries can reflect the expected energy level, as the RSSI samples marked with red circles in Fig. 5.3.

To investigate the effect of τ_{CTC}, we let USRP N210 transmit a set of CTC bits {1, 0, 1, 0, 1, 0}, while τ_{CTC} is set to 160 μs, 128 μs, 96 μs and 64 μs respectively, then collect the RSSI samples at TelosB for each situation. The results are shown in Fig. 5.4. We see that the RSSI distance d_{RSSI}, which is the difference between the maximum and minimum RSSI values, is as high as 9 dB when $\tau_{CTC} = 160$ μs (Fig. 4.10a), and decreases to 5 and 2 dB when $\tau_{CTC} = 128$ μs and $\tau_{CTC} = 96$ μs, respectively. When $\tau_{CTC} = 64$ μs (Fig. 4.10d), the change of RSSI values almost disappears and the CTC bits can not be detected at all.

From the aforementioned analysis, we set τ_{CTC} to 160 μs in this work to make the SLEM performance better.

5.3.1.3 Payload Encoding

Payload encoding is to generate SLEM bits through inserting extra bit to the original WiFi data bits, thus to adjust the expected QAM points in the overlapped subcarriers to deliver CTC information, as shown in Fig. 5.2. The QAM points should have low power when this OFDM symbol is utilized to transmit CTC bit '0', otherwise they should have high power. Each of the low power and high power QAM point has

Table 5.2 An illustration of significance bits under QAM-16

	Low Power				High Power			
	0	0	**0**	**0**	0	0	**1**	**1**
Significance	0	1	**0**	**0**	0	1	**1**	**1**
Bits in Points	1	0	**0**	**0**	1	0	**1**	**1**
	1	1	**0**	**0**	1	1	**1**	**1**
Masks	0	0	1	1	0	0	1	1

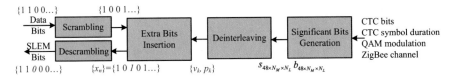

Fig. 5.5 The payload encoding process

two significance bits under QAM-16, as the shadowed ones shown in Table 5.2. Similarly, the point has four and six significance bits under QAM-64 and QAM-256 respectively.

The process of payload encoding is shown in Fig. 5.5. The first problem here is to determine the significant bits for the whole CTC bits, so that they can be used together with the WiFi data bits to determine where and what extra bits should be inserted to generate the SLEM bits. Actually, the significant bits are determined by the CTC bits, the CTC symbol duration τ_{CTC}, QAM modulation, and ZigBee channel which determines the overlapped OFDM subcarriers. We use two matrices S_{48,N_M,N_L} and B_{48,N_M,N_L} to represent the significant bits and the significant bit masks, where 48 is the number of data subcarriers in each OFDM symbol, $N_M = log_2 M$ means there are $log_2 M$ bits for each M-level QAM point, N_L indicates the number of required OFDM symbols for transmitting all the CTC bits. Specifically, $B_{i,j,k} = 1$ if the j-th bit of the QAM point in i-th subcarrier is the significance bit, and the corresponding value of $S_{i,j,k}$ is determined by the expected power level in the k-th OFDM symbol: $S_{i,j,k} = 1$ if the expected power is high, otherwise it is zero. Both matrices S_{48,N_M,N_L} and B_{48,N_M,N_L} will be transformed to one-dimensional arrays $s_{48 \times N_M \times N_L}$ and $b_{48 \times N_M \times N_L}$ respectively, and then passed through deinterleaving as $\bar{s}_{48 \times N_M \times N_L}$ and $\bar{b}_{48 \times N_M \times N_L}$, from which we can easily obtain the significant bits $\{v_k, p_k\}\}$, where v_k and p_k indicate the value and position of the k-th significant bit before deinterleaving, as defined in Sect. 4.3.1.2.

Then, the process of generating the scrambled SLEM bits $\{x_n\}\}$ through inserting extra bits can still follow Algorithm 1. The data stream $\{x_n\}\}$ are finally descrambled to output the SLEM bits. When the SLEM bits are passed through the standard WiFi transmission process shown in Fig. 5.2, both the WiFi and CTC information can be delivered concurrently.

5.3.2 System Design at ZigBee Receiver

In this part, we first introduce the signal classification process with which a ZigBee receiver can differentiate the normal ZigBee and SLEM signals, then we give the detailed design for CTC data detection.

5.3.2.1 Signal Classification

The normal ZigBee transmission utilizes a preamble field in the ZigBee frame to indicate the arrival of a ZigBee signal [6]; thus, the receiver can easily identify a ZigBee signal through this process. Here we use a CTC preamble to indicate the arrival of a SLEM signal at a Zigbee receiver. Then the signal classification process based on the ZigBee preamble and CTC preamble detection is shown in Fig. 5.6. When the received signal energy is over a preset threshold, the ZigBee receiver first conducts ZigBee preamble detection and determines the arrival of a normal ZigBee packet if this preamble is detected, it then uses the standard ZigBee detection to demodulate the ZigBee data bits. If the ZigBee preamble is not detected, the device further conducts CTC preamble detection and determines the arrival of a SLEM signal if this preamble is detected, it then uses CTC data detection process to obtain the CTC bits. We will give detailed design of CTC preamble and its detection in the following parts.

1. CTC Preamble Design
We use the CTC preamble to indicate the arrival of a SLEM signal, and the CTC frame format is depicted in Fig. 5.7. We design the CTC preamble with the fixed pattern of '0101', while the corresponding energy is $\mathbf{E}_{pre} = \{E_L, E_H, E_L, E_H\}$. The first thing we should figure out here is whether this signal pattern can be regarded as normal WiFi or ZigBee signals.

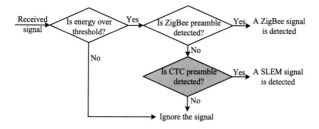

Fig. 5.6 The signal classification process at ZigBee receiver

Fig. 5.7 The CTC frame format

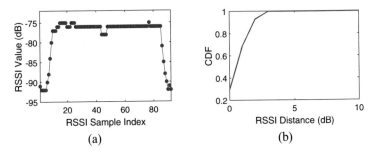

Fig. 5.8 The characteristics of RSSI samples for normal WiFi signals. (**a**) RSSI samples for one WiFi packet. (**b**) CDF of RSSI distances for WiFi packets

The ZigBee signal cannot show this energy pattern as it adopts OQPSK (Offset-QPSK) modulation, which does not change the signal amplitude. Here we mainly investigate the WiFi signal. We make an USRP transmit standard WiFi signals and collect RSSI samples every 32 μs at TelosB. The WiFi and ZigBee central frequencies are 2.472 and 2.475 GHz respectively. Figure 5.8a shows the collected RSSI values of one WiFi packet. We see stable energy levels with a little variation except for the beginning and end of the packet. The main reason comes from the fact that each RSSI value indicate the WiFi signal energy averaged on seven OFDM subcarriers and τ_{CTC} time duration. Considering that the QAM points are randomly distributed in OFDM subcarriers, the averaged signal energy has little change. We further repeat this experiment for about 100 times and calculate the cumulative distributed function (CDF) of RSSI distance d_{RSSI}. Results in Fig. 5.8b show that d_{RSSI} is below 3 dB with the probability of 80%, and it is below 5 dB in nearly all the cases. This analysis indicates that it is almost impossible to detect a CTC preamble from a normal WiFi signal. Therefore, the energy pattern $\mathbf{E}_{pre}$ can be used as the CTC preamble.

2. CTC Preamble Detection
A ZigBee receiver should conduct CTC preamble detection to determine the arrival of a SLEM signal, and this detection is to discerned the energy pattern $\mathbf{E}_{pre}$ from the received signal.

Here we exploit the cross correlation technology, which is always used in the preamble detection process. We let the ZigBee receiver conduct cross correlation between $\mathbf{E}_{pre}$ and the collected RSSI samples $\{r_i\}$. As the values of E_L and E_H vary with the some parameters like the transmission power and the transmitter-receiver distance, we normalize $\mathbf{E}_{pre}$ to a fixed pattern $\{PRE_j\} = \{-1, 1, -1, 1\}(j = 1 \sim 4)$ for calculation. Meanwhile, the number of RSSI samples during one CTC symbol is $N_s = \frac{\tau_{CTC}}{32\mu}$s, making the CTC preamble '0101' with $4 \times N_s$ RSSI samples. Thus, at each position Δ in $\{r_i\}$, the receiver picks up the RSSI samples $\{rp_j^{\Delta}\} = \{r_i, i = \Delta + (j - 1) \cdot N_s\}(j = 1 \sim 4)$ for cross correlation calculation, that

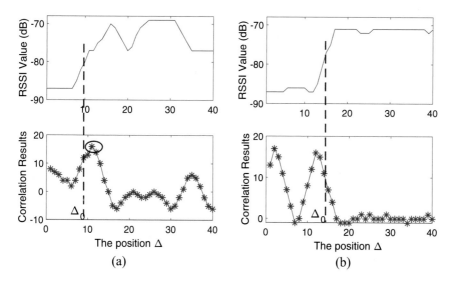

Fig. 5.9 The correlation results of $\{PRE_k\}$ with a SLEM or normal WiFi signal. (**a**) SLEM signal. (**b**) WiFi signal

is, $R_\Delta = \sum_{j=1}^{4} rp_j^\Delta \cdot PRE_j$. When a CTC preamble is received, and the energy pattern of $\{rp_j^\Delta\}$ will match $\mathbf{E}_{pre}$, leading to a peak value in the correlation results.

We further conduct experiments to investigate the feasibility of CTC preamble detection. We let USRP transmit normal WiFi signals and SLEM signals. The SLEM signal has format of Fig. 5.7, and τ_{CTC} is set to be 160 μs. The correlation results of $\{PRE_j\}$ with the received signal at each position Δ are shown in Fig. 5.9. We see that from the position Δ_0 where the averaged RSSI value of $\{rp_j^\Delta\}$ is over a threshold (-80 dB in this experiment), the correlation results for the SLEM signal have a peak value, while those for the normal WiFi signal remain small. Through this way, a ZigBee receiver can determine the arrival of a SLEM signal with a high probability.

5.3.2.2 CTC Data Detection

When the CTC preamble is detected, the ZigBee receiver begins to obtain the CTC data bits through energy demodulation.

The receiver should first determine the optimal RSSI sample set from the received RSSI samples to achieve better performance. As shown in Fig. 5.3, the RSSI samples marked with red circles represent the energy of the transmitted CTC symbols best, they are regarded as the optimal RSSI sample set. Actually, this set is easy to obtain after CTC preamble detection process. As shown in Fig. 5.9a, the position where the correlation result has peak value represents the beginning position of the SLEM signal. Since each CTC symbol has the same duration and the CTC preamble duration is fixed, it is easy to calculate the boundary position of each CTC data symbol. The RSSI samples at these positions form the optimal RSSI sample set, which is denoted by $\{\bar{r}_i\}$.

The process of decoding the CTC data bits based on $\{\bar{r}_i\}$ is pretty simple: if an RSSI value is over a threshold β_s, the corresponding bit is '1', otherwise the bit is '0'. As β_s should change with some parameters like the transmission power and transmitter-receiver distance, we use the mean value of $\{r_i\}$ as the threshold.

5.3.3 System Design at WiFi Receiver

The process at the WiFi receiver side is as follows: the receiver first conducts the standard WiFi receiving process to obtain the SLEM bits, then removes the extra bits to get the original WiFi data bits. The positions of extra bits are fixed in the transmit bits, and they are determined by three kinds of information: the ZigBee channel, QAM modulation and coding rate. The latter two information can be obtained directly from the PLCP (physical layer convergence protocol) header of the WiFi packet [5]. The key issue here is to obtain the ZigBee channel. With the SLEM bits, the WiFi receiver can conduct the channel coding and modulation process, then observe the QAM points and determine the ZigBee channel: the QAM points in the overlapped subcarriers are all highest and lowest ones. In the case that the QAM points do not exhibit this feature and they are randomly distributed in the subcarriers, the receiver determines that this signal is a normal WiFi signal and no extra bits should be removed. This process is fully compatible with the 802.11 standard.

5.4 Performance Evaluation

5.4.1 Experimental Settings

We implement a testbed based on USRP N210 and TelosB for performance evaluation. We let one USRP (WiFi Tx) mimic the WiFi transmitter and generate the WiFi signals following IEEE 802.11 standard, while the SLEM bits are generated offline through encoding the original payload according to the CTC data bits, the ZigBee channel and QAM modulation. We use TelosB as the ZigBee receiver to collect the RSSI samples. The USRP operates on the 13th WiFi channel, while the TelosB platforms operate on the 24th ZigBee channel which is overlapped with the WiFi channel. The experiments are conducted in an open space office with the tested background noise of -91 dB.

5.4.2 Performance of CTC Transmission

5.4.2.1 RSSI Distance

Since SLEM uses energy modulation to convey the CTC information, the RSSI distance will obviously affect the CTC performance. In this part, we investigate the change of RSSI distance with the parameters of τ_{CTC}, QAM modulation types and SNR (Signal to Noise Ratio). The first two parameters can be set in the transmitted signal directly, while SNR is obtained through adjusting the USRP transmission power and the distance between USRP and TelosB.

The experimental results in Fig. 5.10 show that higher order QAM modulation leads to larger RSSI distance. For example, when SNR is as high as 23 dB (Fig. 5.10c), the RSSI distances under QAM-256, QAM-64 and QAM-16 are 23, 16, and 9 dB respectively. Meanwhile, lower τ_{CTC} and SNR lead to shorter RSSI distance, such as Fig. 5.10a which shows that the RSSI distance under $\tau_{CTC} = 160\,\mu s$ is about 5 dB larger than that under $\tau_{CTC} = 96\,\mu s$ when SNR is 23 dB.

5.4.2.2 CTC Preamble Detection

We then conduct experiments to measure the performance of CTC preamble detection, which is designed to determine the arrival of a SLEM signal.

According the design in Sect. 5.3.2.1, the CTC preamble detection is to conduct cross correlation between $\{PRE_j\}$ with the received RSSI samples, and determine the arrival of a SLEM signal if the correlation result R_Δ is higher than a threshold β_{corr}. Here we first investigate the typical values of β_{corr}, then measure the performance of CTC preamble detection.

We let USRP N210 transmit CTC packets following the format of Fig. 5.7. Figure 5.11 shows the correlation results at 15 continuous positions when the average energy is over -80 dB, under QAM-64 and two SNR situations. The high SNR is 25 dB and the low SNR is 8 dB. The correlation results have higher values with higher SNR and larger τ_{CTC}, and β_{corr} should be set considering the cases under different parameters, so we let $\beta_{corr} = 8$ for QAM-64. The values under QAM-16 and QAM-256 can also be set in the same way.

We then use the metric of detection ratio to evaluate the performance of CTC preamble detection, as shown in Fig. 5.12. We see that the detection ratio is as high as 100% when SNR is over 18 dB, while the ratio decreases dramatically under low SNR situations. However, QAM-256 still exhibits high performance when SNR is only 8 dB, especially under $\tau = 160\,\mu s$. We note that the performance of QAM-16 under $\tau = 96\,\mu s$ is not shown here as it is very poor.

Fig. 5.10 The RSSI distance in terms of SNR and τ_{CTC} under different QAM modulation types. (**a**) QAM-16. (**b**) QAM-64. (**c**) QAM-256

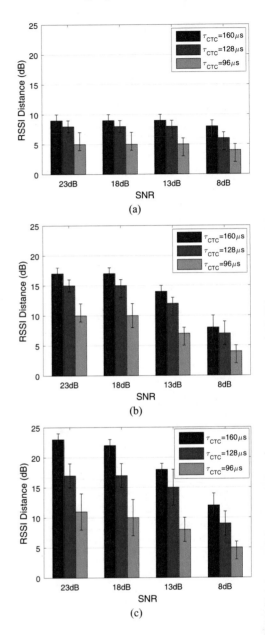

5.4.2.3 SER and PER

We finally use the metrics of SER (symbol error rate) and PER(packet error rate) to investigate the performance of CTC data detection. The experimental settings are the same as those in the previous part. Figure 5.13 shows SER of CTC transmission

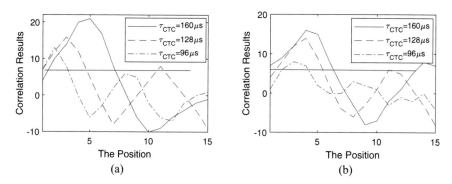

Fig. 5.11 The correlation results of the CTC preamble under QAM-64. (**a**) High SNR. (**b**) Low SNR

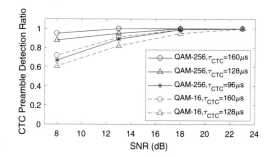

Fig. 5.12 The CTC preamble detection ratio under different situations

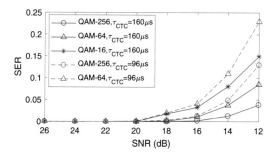

Fig. 5.13 SER of CTC transmission in terms of SNR under different modulation types and τ_{CTC}

in terms of SNR under different QAM modulation types and τ_{CTC} values. We see that SER is as low as zero nearly in all cases when SNR is above 18 dB, and it increases dramatically when SNR decreases from 18 dB. Meanwhile, smaller τ_{CTC} results in more errors, and SER reaches 15% when $\tau_{CTC} = 96\,\mu s$ and SNR of 12 dB even under QAM-256. We also see that the combination of QAM-256 and $\tau_{CTC} = 96\,\mu s$ results in a better performance than the combination of QAM-16 and

Fig. 5.14 PER of CTC
transmission in terms of SNR
under different modulation
types and τ_{CTC}

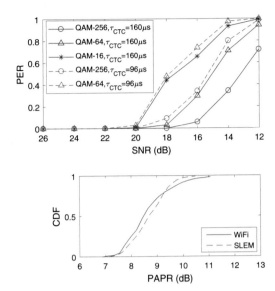

Fig. 5.15 The cumulative
distribution function (CDF)
of PAPR for both the normal
WiFi and SLEM signals
under QAM-16

$\tau_{CTC} = 160\,\mu s$. These results can be used as an important reference when devices select parameters for CTC transmission.

Figure 5.14 depicts the PER of CTC transmission in each situation of Fig. 5.13, and the CTC packet length is 32 bits. We see that PER is as low as zero when SNR is over 20 dB in all the situations. As SNR decreases to 16 dB, PER is still below 0.1 when $\tau_{CTC} = 160\,\mu s$ and QAM-256 is adopted, but exhibits very poor performance under other situations.

5.4.3 Performance of WiFi Transmission

We finally investigate the impact of SLEM design to WiFi transmission.

One key related characteristic which affects the WiFi performance is the peak-to-average-ratio (PAPR) of the time domain signal, as the higher PAPR results in lower performance due to degrading the efficiency of the power amplifier, thus may lead to lower transmission power with the same transmission gain. We get the cumulative distribution function (CDF) of PAPR for the two kinds of signals, the results are shown in Fig. 5.15. We see that the PAPR of SLEM signal looks similar with that of the WiFi signal. We also test the receiving power levels of the two kinds of signals under the same configurations, such as the transmission gain and transmitter-receiver distance, and find that they have no distinguishable difference. These results show that the SLEM design has little effect on the WiFi signal transmissions except the slightly decreased data rate.

5.5 Summary

In this chapter, we present the design and implementation of SLEM, a CTC method which delivers both the WiFi and CTC data bits concurrently through one standard WiFi packet. The current mechanism mainly focuses on the physical layer design. Since CTC can be used to exchange coordination information among heterogeneous wireless devices, it can surely benefit cross-technology interference management with protocol design in the upper layer, which is worthy of further study.

References

1. Chebrolu, K., Dhekne, A.: Esense: communication through energy sensing. In: Proceedings of the ACM MobiCom (2009)
2. Cho, H.-W., Shin, K.G.: BlueFi: Bluetooth over WiFi. In: Proceedings of the ACM SIGCOMM (2021)
3. Gast, M.S.: 802.11ac: A survival guide. O'Reilly Media, Sebastopol (2013)
4. Guo, X., Zheng, X., He, Y.: WiZig: cross-technology energy communication over a noisy channel. In: Proceedings of the IEEE INFOCOM (2017)
5. IEEE Computer Society. 802.11: Wireless LAN medium access control (MAC) and physical layer (PHY) specifications – Amendment 5: enhancements for higher throughput (2009)
6. IEEE Computer Society. 802.15.4: IEEE standard for low-rate wireless networks (2015)
7. Jiang, W., Yin, Z., Kim, S.M., He, T.: Transparent cross-technology communication over data traffic. In: Proceedings of the IEEE INFOCOM (2017)
8. Kim, S.M., He, T.: FreeBee: cross-technology communication via free side-channel. In: Proceedings of the ACM MobiCom (2015)
9. Li, Z., He, T.: WEBee: physical-layer cross-technology communication via emulation. In: Proceedings of the ACM MobiCom (2017)
10. Yao, J., Zheng, X., Xie, R., Wu, K.: Cross-technology communication for heterogeneous wireless devices through symbol-level energy modulation. IEEE Trans. Mobile Comput. 21(11), 3926–3940 (2022)
11. Yin, Z., Jiang, W., Kim, S.M., He, T.: C-Morse: cross-technology communication with transparent Morse coding. In: Proceedings of the IEEE INFOCOM (2017)
12. Yin, Z., Li, Z., Kim, S.M., He, T.: Explicit channel coordination via cross-technology communication. In: Proceedings of the ACM MobiSys (2018)
13. Zhang, Y., Li, Q.: HoWiES: a holistic approach to ZigBee assisted WiFi energy savings in mobile devices. In: Proceedings of the IEEE INFOCOM (2013)
14. Zhang, X., Shin, K.G.: Gap sense: lightweight coordination of heterogeneous wireless devices. In: Proceedings of the IEEE INFOCOM (2013)
15. Zheng, X., He, Y., Guo, X.: StripComm: interference-resilient cross-technology communication in coexisting environments. In: Proceedings of the IEEE INFOCOM (2018)

Chapter 6
Conclusion and Future Work

Abstract In this chapter, we finally summarize this book and put forward several future research directions.

Keywords Cross-technology interference management · Cross-technology communication

6.1 Conclusion

In this book, we introduce the background of heterogeneous wireless networks and focus on the cross-technology coexistence problem which affect the network performance. Considering the differences in channel, physical and MAC layer mechanisms of heterogeneous wireless technologies, we present a series of protocol design to improve the coexistence efficiency, including heterogeneous signal identification, CTI management and CTC protocol design.

In Chap. 2, we propose two signal identification methods, ZShark-FFT which utilizes features extracted through FFT (Fast Fourier Transform), and ZShark- CNN which utilizes CNN (Convolutional Neural Network), for fast heterogeneous signal identification. We evaluate their performance on dataset constructed from collected signals through hardware testbed. Based on these methods, we introduce two CIT management protocols and one CTC protocol to improve the efficiency of cross-technology coexistence.

In Chap. 3, we propose E-CCA, a time-domain protocol design to mitigate CTI in heterogeneous wireless networks. E-CCA utilizes ZShark proposed in Chap. 2 to identify the signal type within several microseconds. It contains an enhanced CCA design, with which a WiFi device can still determine the channel to be busy when the detected energy is below the CCA threshold while a ZigBee signal is identified, avoiding CTI to ZigBee transmissions. Simulations based on NS-3 show that E-CCA can improve ZigBee network performance dramatically, with little sacrifice to WiFi performance.

In Chap. 4, we propose SledZig, a frequency-domain protocol design to mitigate CTI in heterogeneous wireless networks. SledZig utilizes ZShark proposed in Chap. 2 to identify the ZigBee channel, it then decreases the WiFi signal power on the ZigBee channel through making constellation points in the overlapped subcarriers with the lowest power. It can be achieved through encoding the WiFi payload to generate the transmit bits; when the transmit bits are passed through the WiFi transmission process, the signal power on the ZigBee channel can be decreased naturally. We implement and evaluate this protocol on hardware testbed, and experimental results show that it can effectively increase ZigBee transmissions and improve its performance over a WiFi channel with little WiFi throughput loss.

In Chap. 5, we propose SLEM, a CTC method through symbol-level energy modulation. SLEM utilizes ZShark proposed in Chap. 2 to identify the ZigBee channel. It adjusts the QAM points on subcarriers overlapped with the ZigBee channel to make each symbol have distinguishable low or high power levels, which can be achieved through inserting extra bits to the WiFi payload. SLEM can deliver both the WiFi and CTC data bits concurrently through one standard WiFi packet. We implement and evaluate SLEM on hardware testbed, and experimental results show that it can achieve a robust and fast concurrent transmissions of CTC and WiFi.

6.2   Future Research Directions

As more technologies are deployed in wireless networks, such as LoRa [5] which can also operate in the same ISM band, the cross-technology coexistence will become more complex. Meanwhile, the study of ISAC (integration of sensing and communication) [3] in recent years also brings us more inspiration for protocol design. We believe the following directions are worthy of further study.

(1) Coordinated Mechanism Design for Interference Management
The current commercial devices in wireless networks adopt CSMA/CA for channel access, an uncoordinated interference management mechanism as each device makes independent channel access decisions based on the detected signal power. However, this method has relatively low efficiency comparing with the coordinated mechanism, which is accomplished by exchanging coordination information between heterogeneous nodes. In particular, the emergence of CTC has brought great convenience to the coordinated interference management design. Actually, there are already some related works, such as ECC [4] which makes a WiFi AP coordinate data transmissions of all the WiFi and ZigBee devices to avoid interference. However, the current research is limited to two or three kinds of wireless networks, like WiFi and ZigBee [2, 4]. When considering all the coexisted networks, it will be more difficult to achieve effective interference management. The main challenges include how to design control frames for transmission between multiple heterogeneous devices, how to reduce the transmission overhead, and how to make the protocol design compatible to the standard.

(2) Sensing-Assisted Cross-Technology Coexistence Design

To improve the cross-technology coexistence efficiency, now we only focus on the network protocol or communication mechanism design. However, with the study of ISAC, we believe that a variety of sensing information can be used to assist the coexistence design. For example, the signal type and channel are sensed through the signal identification method and can be used for CTI management and CTC design to improve the coexistence efficiency in this book. However, we believe that current IoT devices can perceive much richer information, such as data collected by cameras and other different sensors, geographic environment information, etc. There have already been some relevant studies, such as [1] which utilizes the geographical environment information to automatically estimate the LoRa link quality, so as to obtain better network performance. We believe these sensing data can also be used in cross-technology coexistence scenario to further improve the coexistence efficiency.

(3) Waveform Design Through Payload Encoding

Waveform design is very important in ISAC to make one wireless device have the ability of simultaneous sensing a target and communicating with another device. There have already been some research efforts on theoretical waveform designs to improve the performance of sensing and communication [6]. However, these designs require hardware modifications to the transceiver, which cannot be implemented on the commercial devices. Considering that there are a large number of commercial devices in real networks, the cost of deploying ISAC through hardware updating is unaffordable. In this book, we have already investigated payload encoding for controlling the signal waveform, thus to achieve CTC and CTI management. We can further study the feasibility of using payload encoding for ISAC waveform design. In this way, ISAC functions can be realized on commercial devices at the expense of some transmission performance, and the deployment cost can be saved significantly.

References

1. Demetri, S., Zuniga, M., Picco, G.P., Kuipers, F., Bruzzone, L., Telkamp, T.: Automated estimation of link quality for LoRa: a remote sensing approach. In: Procedings of the ACM IPSN (2019)
2. Li, Y., Chi, Z., Liu, X., Zhu, T.: Chiron: concurrent high throughput communication for IoT devices.. In: Proceedings of the ACM MobiSys (2018)
3. Tan, D., He, J., Li, Y., Bayesteh, A., Chen, Y., Zhu, P., Tong, W.: Integrated sensing and communication in 6G: motivations, use cases, requirements, challenges and future directions. In: Proceedings of the IEEE JC&S (2021)
4. Yin, Z., Li, Z., Kim, S.M., He, T.: Explicit channel coordination via cross-technology communication. In: Proceedings of the ACM MobiSys (2018)
5. Yu, F., Zheng, X., Liu, L., Ma, H.: LoRadar: an efficient LoRa channel occupancy acquirer based on cross-channel scanning. In: Proceedings of the IEEE INFOCOM (2022)
6. Zhou, W., Zhang, R., Chen, G., Wu, W.: Integrated sensing and communication waveform design: a survey. IEEE Open J. Commun. Soc. 3(1), 1930–1949 (2022)

Printed in the United States
by Baker & Taylor Publisher Services